Computational Fluid Dynamics

Computational Fluid Dynamics

Proceedings of the Fourth UNAM Supercomputing Conference

Mexico City, Mexico   27 – 30 June 2000

Editors

Eduardo Ramos
*Centro de Investigación en Energía,
Universidad Nacional Autónoma de México*

Gerardo Cisneros
SGI (Silicon Graphics, S. A. de C. V.)

Rafael Fernández-Flores
Alfredo Santillán-González
*Dirección General de Servicios de Cómputo Académico,
Universidad Nacional Autónoma de México*

World Scientific
New Jersey • London • Singapore • Hong Kong

Published by

World Scientific Publishing Co. Pte. Ltd.

P O Box 128, Farrer Road, Singapore 912805

USA office: Suite 1B, 1060 Main Street, River Edge, NJ 07661

UK office: 57 Shelton Street, Covent Garden, London WC2H 9HE

British Library Cataloguing-in-Publication Data
A catalogue record for this book is available from the British Library.

COMPUTATIONAL FLUID DYNAMICS
Proceedings of the IV UNAM Supercomputing Conference

ISBN 981-02-4535-1

Printed in Singapore by Mainland Press

Preface

Numerical solution of the governing equations of fluid dynamics has been both a challenge and a source of useful information in the last fifty years since the first electronic computers became available.

The lack of analytical tools and the pressing need to describe and predict various astrophysical and geophysical flows and solve many engineering problems, prompted scientists and engineers to develop methods to solve approximate versions of the fluid dynamics equations. Early attempts were unsuccessful due to incomplete theories of errors and the impossibility to perform very large amount of floating point operations in reasonable amounts of time.

These two topics of scientific activity have seen tremendous progress in the last years; digital electronic computers have increased in speed and memory capacity by many orders of magnitude. Presently, solutions to Navier–Stokes equations have been obtained for as many as $2000^3 = 8 \times 10^9$ control volumes in clusters of supercomputers as described in the paper authored by Woodward *et al.* in the present volume. Another promising avenue for finding the solution for problems requiring intensive numerical calculations is the parallel solution of independent segments of the total volume of integration. Although these methods are still in the process of being refined, they have proved their usefulnes in several important examples of transfer processes in engineering and meteorology. Details of such methods are discussed in the papers by Carey *et al.* and Jabouille.

The recent advances in various areas of computational fluid dynamics have transformed the field from a purely academic activity into an intensive topic for scientific research and a useful tool for engineering design. Given its importance in many fields, it was considered appropiate to organize a meeting where the latest developments in the area were presented by leading specialists. With this purpose in mind, the Fourth UNAM Supercomputing Conference was held on June 27–30, 2000 in *Amoxcalli* (Nahuatl for "house of codices"), the library and conference building of the School of Sciences on the main campus of the National Autonomous University of Mexico (UNAM). This academic event had also as an objective bringing together the international and local communities to boost computational fluid dynamics research in Mexico. Although the original idea was to discuss exclusively fluid dynamics problems that required intensive computing, due to the incipient nature of the Mexican CFD community it was decided to include also contributions that contained interesting and useful scientific information even when the computational requirements were modest. Also, analytical studies that provided natural checks for numerical calculations were presented. A total of 72 participants from 8 counries contributed to generate an intense atmosphere during the four days that the meeting lasted. The scientific program included 8 invited lectures, 21 contributed talks and 11 poster presentations. The present volume, which contains 27 papers reporting on the work presented at the meeting, is organized as follows. The first part deals with astrophysical gas dynamics while the second part includes papers related to geophysical problems. The third and fourth parts contain papers relating to numerical methods and applications, respectively.

The end product is a collection of research papers that reflect the state of the art of computational fluid dynamics at an international level with emphasis on the work that is being carried out in Mexico. We want to thank the members of the scientific committee who reviewed and selected the abstracts for inclusion in the conference program and refereed most of the full papers. Also, we express our gratitude to the other anonymous reviewers who selflessly read and commented on the full papers. Dr. Fernando Magaña, director of the School of Sciences of the UNAM kindly allowed us to use the Carlos Graef auditorium in *Amoxcalli*. We appreciate the assistance of the administrators of the conference facilities at *Amoxcalli*, Ma. Elena Abrín and Rubén Alba, and the rest of their staff who provided the infrastructure required for the meeting. Mr. Roberto Bonifaz coordinated all supporting services and ran a virtually trouble-free event; it is a pleasure for us to acknowledge his participation in the organizing team. Drs. Víctor Guerra and Geneviève Lucet, respectively Director General and Director of Computing for Research at UNAM's General Directorate of Academic Computing Services (DGSCA) continously supported the meeting organization. We also want to acknowledge the assistance of our budget managers and graphic designers from DGSCA for their competent and efficient work. DGSCA staff volunteered to help us to solve almost all of the small (and not so small) last minute problems. We are especially grateful for the generous grant provided by Silicon Graphics, S.A. de C.V., which made possible the organization of the conference and the production of these proceedings.

Eduardo Ramos
Gerardo Cisneros
Rafael Fernández-Flores
Alfredo Santillán-González
México D.F., México, 2001

Scientific Committee

Rubén Avila
Instituto Nacional de Investigaciones Nucleares, México

Graham F. Carey
Univesity of Texas at Austin, USA

John Kim
University of California, Los Angeles, USA

Douglas Lilly
Univeristy of Oklahoma, USA

Joseph J. Monaghan
Monash University, Australia

Alejandro Raga
Universidad Nacional Autónoma de México (UNAM)

Pierre-Louis Sulem
Observatoire de la Côte d'Azur, France

Paul R. Woodward
University of Minnesota, USA

Organizing Committee

Eduardo Ramos
Centro de Investigación en Energía, UNAM

Roberto Bonifaz
Dirección General de Servicios de Cómputo Académico (DGSCA), UNAM

Gerardo Cisneros-Stoianowski
Silicon Graphics, S.A. de C.V. (SGI)

Rafael Fernández-Flores
DGSCA, UNAM

Alfredo Santillán
DGSCA, UNAM

Enrique Vázquez-Semadeni
Instituto de Astronomía, UNAM

Contents

PART I

ASTROPHYSICS

VERY HIGH RESOLUTION SIMULATIONS
OF COMPRESSIBLE, TURBULENT FLOWS

PAUL R. WOODWARD, DAVID H. PORTER, IGOR SYTINE,
S. E. ANDERSON

Laboratory for Computational Science & Engineering
University of Minnesota, Minneapolis, Minnesota, USA

ARTHUR A. MIRIN, B. C. CURTIS, RONALD H. COHEN, WILLIAM P. DANNEVIK,
ANDRIS M. DIMITS, DONALD E. ELIASON
Lawrence Livermore National Laboratory

KARL-HEINZ WINKLER, STEPHEN W. HODSON
Los Alamos National Laboratory

The steadily increasing power of supercomputing systems is enabling very high resolution simulations of compressible, turbulent flows in the high Reynolds number limit, which is of interest in astrophysics as well as in several other fluid dynamical applications. This paper discusses two such simulations, using grids of up to 8 billion cells. In each type of flow, convergence in a statistical sense is observed as the mesh is refined. The behavior of the convergent sequences indicates how a subgrid-scale model of turbulence could improve the treatment of these flows by high-resolution Euler schemes like PPM. The best resolved case, a simulation of a Richtmyer–Meshkov mixing layer in a shock tube experiment, also points the way toward such a subgrid-scale model. Analysis of the results of that simulation indicates a proportionality relationship between the energy transfer rate from large to small motions and the determinant of the deviatoric symmetric strain as well as the divergence of the velocity for the large-scale field.

1   Introduction

The dramatic improvements in supercomputing power of recent years are making possible simulations of fluid flows on grids of unprecedented size. The need for all this grid resolution is caused by the nearly universal phenomenon of fluid turbulence. Turbulence develops out of shear instabilities, convective instabilities, and Rayleigh–Taylor instabilities, as well as from shock interactions with any of these. In the tremendously high Reynolds number flows that are found in astrophysical situations, turbulence seems simply to be inevitable. Through the forward transfer of energy from large scale motions to small scale ones that characterizes fully developed turbulence, a fluid flow problem that might have seemed simple enough at first glance is made complex and difficult. Because the turbulent motions on small scales can strongly influence the large-scale flow, it is necessary to resolve the turbulence, at least to some reasonable extent, on the computational grid before the computed results converge (in a statistical sense). Thus it is that, more often than not, turbulence drives computational fluid dynamicists to refine their grids whenever increased computing power allows it.   In this paper, we give examples of computations performed recently which illustrate where increased grid resolution is taking us. We focus our attention on the prospect that extremely highly resolved direct numerical simulations of turbulent flows can guide the development of statistical models for representing the effects of turbulent fluid motions that must remain unresolved in computations on smaller grids or of more complex problems.

2   Homogeneous, compressible turbulence

Perhaps the most classic example of a grid-hungry fluid dynamics problem is that of homogeneous, isotropic turbulence. Here we focus all available computational power on a small subdomain that could, in principle, have been extracted from any of a large number of turbulent flows of interest. If from such a simulation we are able to learn the correct statistical properties of compressible turbulence, we can use our data to help test or construct appropriate subgrid-scale models of turbulent motions. When our computational grid is forced to contain an entire large-scale turbulent flow, we can then use such a model to make the computation practical. In the section that follows, we will see an example of such a larger flow in which we have used an 8-billion-cell grid in order to resolve both the large-scale flow and the turbulent fluid motions it sets up. In the results presented in this section, we will encounter signatures of the fully developed turbulence that can be recognized in a variety of such larger, more structured flows.

It is difficult to formulate boundary conditions that correctly represent those for a small subdomain of a larger turbulent flow. We use periodic boundary conditions here, but these are of course highly artificial, and therefore we must be careful not to overinterpret our results. Not only are our boundary conditions problematical, but our initial conditions also raise important issues. We rely on the theoretical expectation that, except for a small number of conserved quantities such as total energy, mass, and momentum, the details of our initial conditions will ultimately be "forgotten" as the turbulence develops, so that they will eventually become irrelevant. After a long time integration, we will find that the behavior of the flow on the scales comparable to the periodic length of our problem domain will remain influenced by both our initial conditions and by our periodic boundary conditions. However, the flow on shorter scales should be characteristic of fully developed turbulence. The flow on the very shortest scales, of course, must be affected by viscous dissipation and numerical discretization errors.

We are interested in the properties of compressible turbulence in the extremely high Reynolds number regime; we have no interest in the effects of viscosity, save upon the steady increase in the entropy of the fluid via the local dissipation of turbulent kinetic energy into heat. Therefore, we use an Euler method, PPM (the

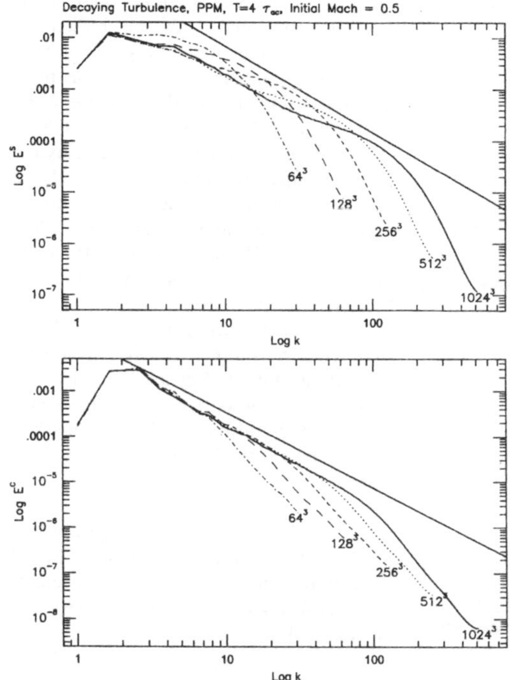

Figure 1. A comparison of velocity power spectra in 5 PPM runs on progressively finer grids, culminating in a grid of a billion cells. All begin with identical initial states and all are shown at the same time, after the turbulence is fully developed. As the grid is refined, more and more of the spectrum converges to a common result. Spectra for the solenoidal (incompressible) component of velocity are shown in the top panel, while the spectra of the compressional component are at the bottom. The straight lines indicated in each panel show the Kolmogorov power law. Note that agreement between the runs extends to considerably higher wavenumbers in the compressional spectra than in the solenoidal ones. This effect, first noted in our earlier work at 512^3 grid resolution, was confirmed in the incompressible limit by simulations by Orszag and collaborators. A similar flattening of the power spectrum just above the dissipation scales has also been observed in data from experiments.

Piecewise-Parabolic Method[1–5]), in order to restrict the effects of viscous dissipation[6] to the smallest range of short length scales that we are able. A similar approach has been adopted by several other investigators (see e.g. [7–10]). We must of course be careful to filter out the smallest-scale motions, which are affected by viscosity and other numerical errors, before we interpret our results as characterizing extremely high Reynolds number turbulence. This approach is in contrast to that adopted by many researchers, who attempt to approximate the behavior of flow in the limit of extremely high Reynolds numbers with the behavior of finite Reynolds number flows, where the Reynolds numbers are only thousands or less. In principle, an Euler computation gives an approximation to the limit of Navier–Stokes flows as the viscosity and thermal conductivity tend to zero. For our gas dynamics flows, this limit should be taken with a constant Prandtl number of unity. Much practical experience over decades of using Euler codes like PPM indicates that this intended convergence is actually realized. However, we must be aware that for turbulent flows, convergence, of course, occurs only in a statistical sense.

We can verify the convergence of our simulation results in this particular case in two ways. First, we can compare results from a series of simulations carried out on different grids. To demonstrate convergence, we look at the velocity power spectra, in Fig. 1, obtained from these flow simulations. We expect that as the grid

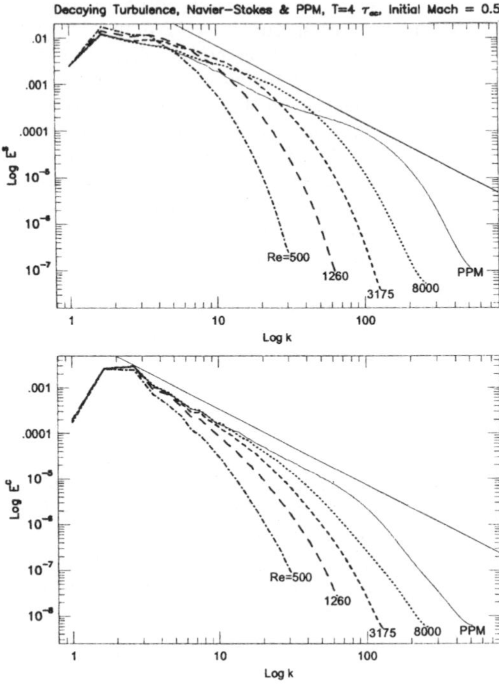

Figure 2. PPM Navier–Stokes simulations are here compared with the billion-cell PPM Euler simulation of Fig. 1. Again, decaying compressible, homogeneous turbulence is being simulated on progressively finer grids of 64^3, 128^3, 256^3, and 512^3 cells. On each grid the smallest Navier–Stokes dissipation coefficients are used that are consistent with an accurate computation. Reynolds numbers are 500, 1260, 3175, 8000. All runs begin with the same initial condition and are shown at the same time, after 4 sound crossings of the principal energy containing scale. As with the PPM Euler runs in Fig. 1, this convergence study (to the infinite Reynolds number limit that we seek) shows that the compressional spectra, in the lower panel, are converged over a longer range in wavenumber than the solenoidal spectra at the top. In fact, the solenoidal spectra display no "inertial range," with Kolmogorov's $k^{-5/3}$ power law, at all. These results and those in Fig. 1 indicate that the Euler spectra are accurate to about 4 times higher wavenumbers than Navier–Stokes.

is successively refined, the velocity power spectrum on large scales does not change, while that on small scales is altered. The part of the spectrum that does not change on each successive grid refinement should, ideally, extend to twice the wavenumber each time the grid is refined. That this behavior is in fact observed is shown in Fig. 1, taken from [11].

We would also like to verify that our procedure not only converges, but that it converges to the high Reynolds number limit of viscous flows. We can do this by comparing Navier–Stokes simulations of this same problem, carried out on a series of successively refined grids, with our PPM Euler simulation on the finest, billion-cell grid. This comparison is shown in Fig. 2, also taken from [11]. (A similar comparison for 2D turbulence is given in [12].) The Navier–Stokes simulations do not achieve sufficiently high Reynolds numbers to unquestionably establish that they are converging to the same limit solution as the Euler runs. This is because the 512^3 grid of the finest Navier–Stokes run has only a Reynolds number of 8000. This Reynolds number has been limited by our demand that each Navier–Stokes simulation represent a run in which the velocity power spectrum has converged. This convergence has been checked, on the coarser grids where this can be done, by refining the grid while keeping the coefficients of viscosity and thermal conductivity constant and by verifying that the velocity power spectrum agrees over the entire range possible. We are confident that, had we been able to afford to carry the sequence of Navier–Stokes simulations forward to grids of 2048^3 or 4096^3 cells, we would have been able to obtain strict agreement with the already converged portion of the PPM velocity power spectrum on the billion-cell grid. At this time, such a demonstration is not practical.

We can use the detailed data from the billion-cell PPM turbulence simulation to test the efficacy of proposed subgrid-scale turbulence models. We can filter out the shortest scales affected either directly (from wavelengths of 2 Δx to about 8 Δx) or indirectly (from about 8 Δx to about 32 Δx) by the numerical dissipation of the PPM scheme. We are then left with the energy-containing modes, from wavelengths of about 256 Δx to 1024 Δx, and the turbulent motions that these induce, from about 256 Δx to about 32 Δx. A large eddy simulation, or LES, would involve direct numerical computation of these long wavelength disturbances with statistical "subgrid-scale" modeling of the turbulence. We can test a subgrid-scale turbulence model with this data by comparing the results it produces, in a statistical fashion, with those which our direct computation on the billion-cell grid has produced in the wavelength range between 256 and 32 Δx. If we were to identify a subgrid-scale model that could perform an adequate job, as measured by the above procedure, then we should be able to add it to our PPM Euler scheme on the billion-cell grid. We will discuss aspects of such a model in the next section. Such a statistical turbulence model should, in the case of PPM, not be applied at the grid scale, as is generally advocated in the turbulence community, but instead at the scale where PPM's numerical viscosity begins to damp turbulent motions. This scale is about 8 Δx , as can easily be verified by direct PPM simulation of individual eddies, as in [6], or by examination of the power spectra in Fig. 1 (see especially the power spectra for the compressible modes).

In earlier articles (see for example [13]) we have suggested that the flattening of the velocity power spectra just before the dissipation range, seen in both PPM and Navier–Stokes simulations (see Figs. 1 and 2), is the result of diminished forward transfer of energy to smaller scales. This diminished forward energy transfer is caused by the lack of such smaller scales in the flow as a result of the action of the viscous dissipation. Applying an eddy viscosity from a statistical model of turbulence on scales around 8 Δx in a PPM turbulence simulation should, if the eddy viscosity has the proper strength, alleviate the above mentioned distortion of the forward energy transfer and make the simulated motions more correct in the range from 32 Δx to 8 Δx , where we observe the flattening of the power spectra for the solenoidal modes in PPM Euler calculations. If the turbulence model, under the appropriate conditions, produces a negative viscous coefficient, this should help to give the PPM simulated flow the slight kick needed for it to develop turbulent motions on the scales resolved by the grid in this same region from about 32 Δx to about 8 Δx. Thus we can hope that a successful statistical model of turbulence would make the computed results of such an LES computation with PPM accurate right down to the dissipation range at wavelengths of about 8 Δx for both the

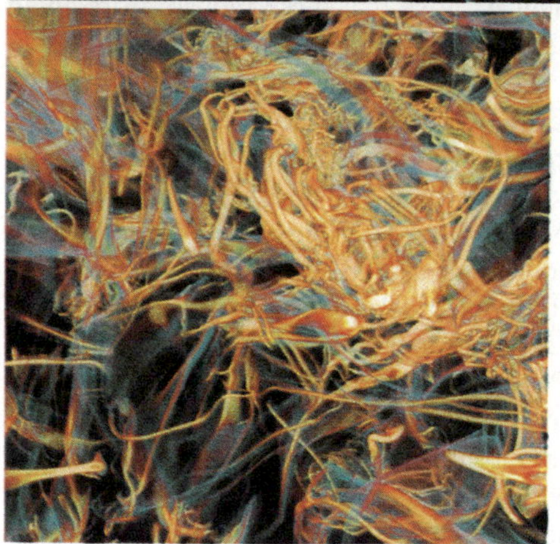

Figure 3. Four snapshots of the distribution of the magnitude of vorticity in a region of the billion-cell PPM simulation, showing stages in the transition to fully developed turbulence. The smallest vortex tube structures are in the dissipation range and have diameters of roughly 5 grid cells. The transition to turbulence, believe it or not, is not quite complete at the time of the final picture of this sequence.

compressible and the solenoidal components of the velocity field. We note that PPM Euler computations enjoy this accuracy in nonturbulent flow regions. The desired effect of an LES formulation of PPM would be that this level of simulation accuracy would be maintained for the solenoidal velocity field within turbulent regions as well. Such an LES formulation should improve the ability of the numerical scheme to compute correct flow behavior, so that in these regions it would match the results of a PPM Euler computation on a grid refined by a factor between 2 and 4 in each spatial dimension and time. The enhanced resolving power in these regions of such a PPM LES scheme over an accurate Navier–Stokes simulation at the highest Reynolds number permitted by the grid would be very much greater still, as the power spectra in Figure 2 clearly indicate. In this statement, we have of course assumed that an approximation to the infinite Reynolds number limit is desired, and not a simulation of the flow at any attainable finite Reynolds number. In astrophysical calculations, as in many other circumstances, this is generally the case.

In addition to providing the essential, highly resolved simulation data needed to validate subgrid-scale turbulence models for use in our PPM scheme, our billion-cell PPM Euler simulation of homogeneous, compressible turbulence gives a fascinating glimpse at the process of transition to fully developed turbulence. In the sequence of snapshots of the distribution of the magnitude of vorticity in this flow given in Fig. 3, we see that the vortex sheet structures that emerged from our random stirring of the flow on very large scales develop concentrations of vorticity in ropes that become vortex tubes. These vortex tubes in turn entwine about each other as the flow becomes entirely turbulent. We note that our first billion-cell simulation of this type was performed in collaboration with Silicon Graphics, who in 1993 built a prototype cluster of multiprocessor machines expressly for attacking such very large computational challenges. The simulation shown in Figure 3 was performed in 1997 on a cluster of Origin 2000 machines from SGI at the Los Alamos National Laboratory. Although we have used results of this simulation to test subgrid-scale turbulence model concepts, we will discuss detailed ideas for such models in the context of an even more highly resolved flow.

3   Turbulent fluid mixing at an unstably accelerated interface

An example of a turbulent flow driven by a large-scale physical mechanism is that of the unstable shock-acceleration of a contact discontinuity (a sudden jump in gas density) in a gas. This calculation was carried out as part of the DoE ASCI program's verification and validation activity, and it was intended to simulate a shock tube experiment of Vetter and Sturtevant[14] at Caltech. In the laboratory experiment, air and sulfur hexafluoride were separated by a membrane in a shock tube, and a Mach 1.5 shock impinged upon this membrane, forcing it through an adjacent wire mesh and rupturing it. The interface between the two gases is unstable when accelerated by a shock, and both the large-scale flexing of the membrane and the wire mesh impart perturbations that are amplified by this Richtmyer–Meshkov instability. The conditions of this problem therefore provide a context to observe the competition and interaction of small- and large-scale perturbations of the interface along with the turbulence that develops.

In Fig. 4, at the top left on the next page, a thin slice through the unstable mixing region between the two gases is shown in a volume rendering of the entropy of the gas. The entropy, after the initial shock passage, is a constant of the motion, with different values in each gas. In the figure, white corresponds to the entropy of the pure initially denser gas, while the pure initially more diffuse gas is made transparent. The regions of intermediate colors in the figure show different proportions of mixing of the two fluids within the individual cells of the 1920×2048^2 grid (8 billion cells). Below this image, a volume rendering of the enstrophy, the square of the vorticity, is shown for the same slice. This simulation was carried out on the Lawrence Livermore National Laboratory's large ASCI IBM SP system using 3904 CPUs. A constraint of this particular supercomputing opportunity was that the previously tuned simplified version, sPPM[5], of our PPM[1-4] gas dynamics code had to be used. This constraint limited us to a single-fluid model with a single gamma-law equation of state to simulate the air and sulfur hexafluoride system in the Caltech experiment. We chose a gamma value of 1.3 and initial densities of 1.0 and 4.88 to represent the experiment as best we could under these constraints. The interface perturbation was $0.01 \times [-\cos(2\pi x)\cos(2\pi y) + |\sin(10\pi x)\sin(10\pi y)|]$.

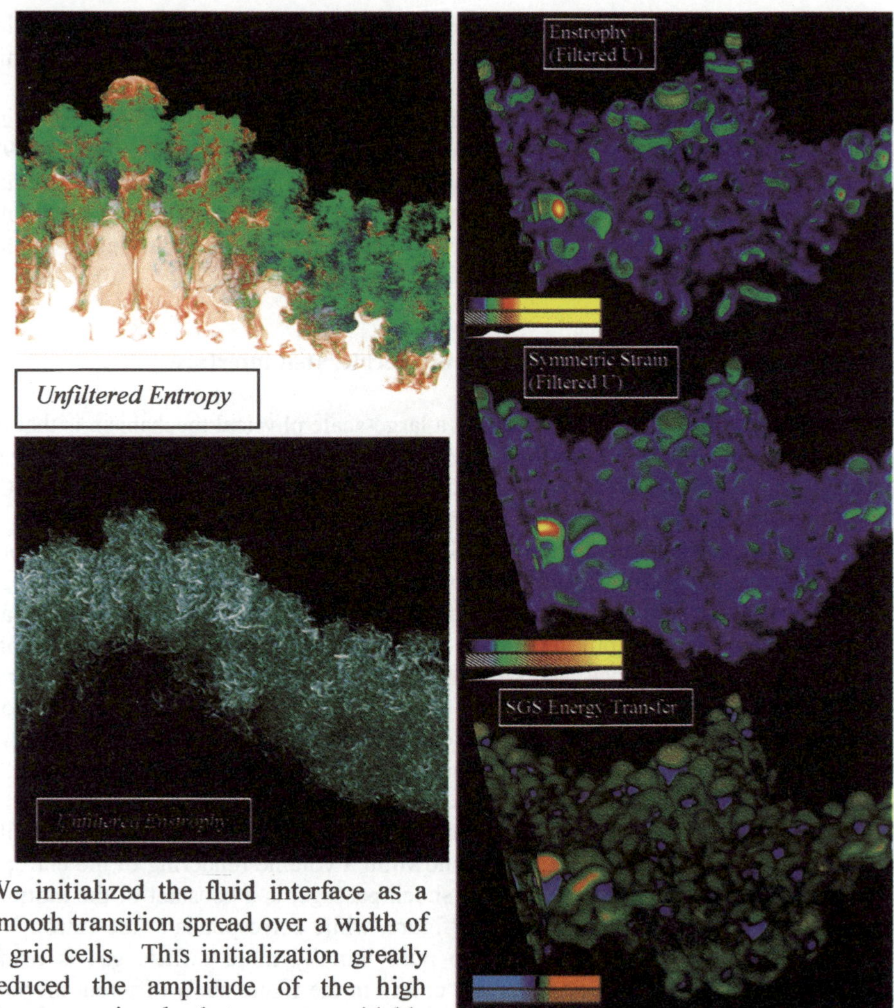

We initialized the fluid interface as a smooth transition spread over a width of 5 grid cells. This initialization greatly reduced the amplitude of the high frequency signals that are unavoidable in any grid-based method. The sPPM method of capturing and advecting fluid interfaces forces smearing of these transitions over about 2 grid cells and resists, through its inherent numerical diffusion, development of very short wavelength perturbations. By setting up the initial interface so smoothly, we assured that after its shock compression it would contain only short wavelength perturbations that the sPPM scheme was designed to handle. Nevertheless, the flow is unstable, so one must be careful in interpreting the results.

A detailed discussion of the results of this simulation will be presented elsewhere[15]. As with the homogeneous turbulence simulations discussed earlier and our simulations of compressible convection in stars[6,16,17], this Richtmyer–Meshkov problem demonstrates convergence upon mesh refinement to a velocity power spectrum, in this case for the longitudinal velocity, shown in the figure to the right on this page, that demonstrates an energy-containing range determined by the initial and boundary conditions of the problem with a short inertial range with Kolmogorov $k^{-5/3}$ scaling. At the

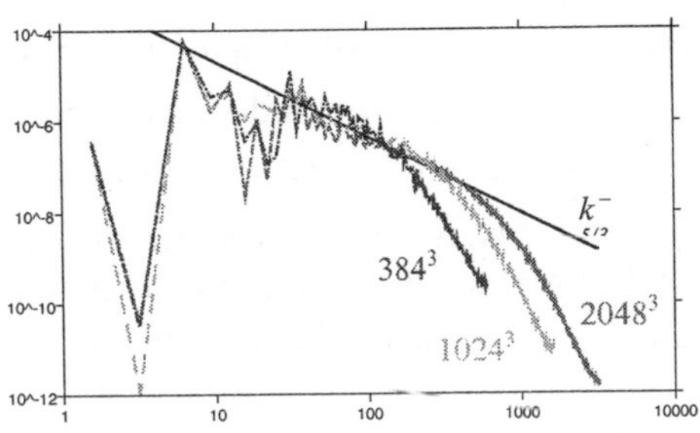

Power spectra for the longitudinal velocity at the midplane in 3 Richtmyer–Meshkov simulations with sPPM on progressively finer grids.

highest wavenumbers, the numerical dissipation of the sPPM Euler scheme is at work, and there is again, just above this dissipation range, a short segment of the spectrum with a slope flatter than the Kolmogorov trend. The 8-billion-cell grid of this calculation is so fine that it allows us to apply a Gaussian filter with full width at half maximum of 67.8 cells, producing a complex filtered flow consisting mostly of the energy-containing modes (a 128-cell sine wave is damped by a factor of $1/e$). We interpret the many well-resolved modes at scales removed by the filter as fully developed turbulence. Using the PPM code, we might hope to capture the modes of the filtered fields on a grid of 256^3 cells. If we can use our 8-billion-cell data to characterize the statistical effects of the turbulent modes beneath the 128-cell scales preserved by the filter, then an LES calculation on a very much coarser grid that made use of this characterization might succeed in producing the correct statistically averaged behavior of the mixing layer. With this goal in mind, we consider the rate, F_{SGS}, of forward energy transfer from the modes preserved by the filter to those eliminated by it.

If we denote the filtered value of a variable, such as the density ρ, by an overbar, $\overline{\rho}$, then we will denote by a tilde the result of a mass-weighted filtering, as for the velocity: $\widetilde{u} = \overline{\rho u} / \overline{\rho}$. With these definitions, the filter when applied to the equation of momentum conservation produces the following result:

$$\frac{\partial \overline{\rho}\widetilde{u}_i}{\partial t} + \partial_j (\overline{\rho}\widetilde{u}_i\widetilde{u}_j) = \partial_i \overline{p} + \partial_j \tau_{ij}$$

where $\tau_{ij} = \overline{\rho u_i u_j} - \overline{\rho}\widetilde{u}_i\widetilde{u}_j$ is a quantity that we call the "subgrid-scale" stress (considering, for our present purposes, the computed structures below the filter scale to be "subgrid-scale" structures in an imagined LES computation). The filtered kinetic energy is now $\widetilde{K} = \overline{\rho}\widetilde{u}^2 / 2$, and the equation for $\partial \widetilde{K} / \partial t$ that can be derived from the momentum equation above and the continuity equation contains a term $-\widetilde{u}_i \partial_j \tau_{ij}$, which is in turn $-\partial_j(\widetilde{u}_i \tau_{ij}) + (\partial_j \widetilde{u}_i)\tau_{ij}$. The first term in this second expression is the divergence of an energy flux, and the negative of the second term, $-(\partial_j \widetilde{u}_i)\tau_{ij}$, we identify as the rate of kinetic energy transfer, F_{SGS}, from the modes on scales larger than our filter to the "subgrid-scale" modes on smaller scales. In the illustrations two pages earlier, this forward energy transfer rate, F_{SGS}, is visualized along with the enstrophy (square of the vorticity for $\widetilde{u}$) and

the deviatoric symmetric strain, $$\Pi^2 = \sum_{ij}\left(\frac{\partial \widetilde{u}_i}{\partial x_j} + \frac{\partial \widetilde{u}_j}{\partial x_i} - \frac{2}{3}\delta_{ij} \nabla \cdot \vec{u}\right)^2.$$

Although there is a positive correlation between F_{SGS} and Π^2, it is clear from these images that the forward energy transfer rate is both positive and negative. Thus if we use Π^2 to model F_{SGS}, we must include another factor which switches sign at the appropriate places in the flow. The situation is clearest in the region of the largest plume in the center of the problem domain. The "mushroom cap" at the top of this plume is essentially a large ring vortex, as indicated in the diagram at the lower right. Near the top of the plume, the top in the diagram, there is an approximate stagnation flow, with compression in the direction along the plume and with expansion in the two dimensions of the plane perpendicular to this, the plane of the original unstable layer. The transfer of energy to small scales is large in this region. Here we believe that small-scale line vortices are stretched so that they tend to become aligned in this plane in myriad directions, which leads to their mutual disruption to produce even smaller-scale line vortex structures. We have observed this process in great detail at the tops of rising, buoyant plumes in our earlier high-resolution simulations of stellar convection (cf. [17,18] and particularly the movie looking down on the top of the simulated convective layer). At the base of the vortex ring in the diagram at the right, and in the lower portion of the "mushroom cap" of the large, central plume in the simulated Richtmyer–Meshkov mixing layer, the flow compresses in two dimensions while it expands in the third, the dimension along the length of the plume. Here there is energy transfer from the small scales to the large, as indicated by negative values of F_{SGS} in the volume-rendered image. Here we believe that vortex tubes become aligned in the single stretching direction, so that they are likely to interact by entwining themselves around each other to form larger vortex tube structures, a

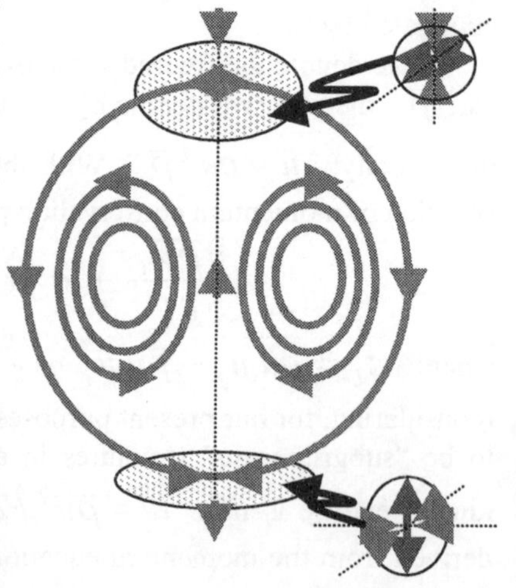

process that we have also observed in greater detail in our stellar convection studies (cf. [17,18] and the portion of the movie that shows the conglomeration of vertically aligned vortex tubes in the downflow lanes along the edges of the convection cells). This behavior of the forward energy transfer rate, F_{SGS}, with its sign dependence on the nature of the local flow field, gives us the hint that this transfer rate, which must be central to any successful subgrid-scale model of turbulence, should be modeled in terms of the determinant,

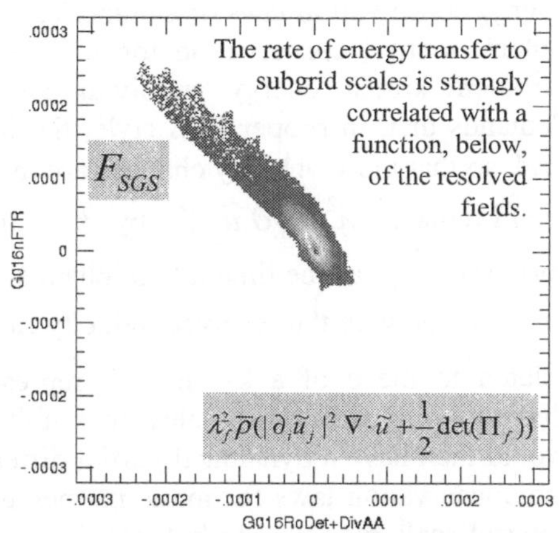

$\det(\Pi)$, of the deviatoric symmetric strain tensor for the filtered velocity field, which flips sign in the appropriate locations (since the determinant is just the product of the 3 eigenvalues of the matrix). This yields a much better correlation than using Π^2, as is the more usual choice. The correlation is improved still further by including the obvious dependence of F_{SGS} on the divergence of the filtered velocity field, which transports turbulent kinetic energy from larger to smaller scales by simply compressing the overall flow in a region. Thus we obtain:

$$F_{SGS} \propto -\lambda_f^2 \overline{\rho} \ (|\partial_i \tilde{u}_j|^2 \nabla \cdot \tilde{u} + \frac{1}{2} \det \Pi_f)$$

where Π_f represents Π for the filtered velocity field, and λ_f is a filter wavelength, equal to 128 cells for our Gaussian filter with full width at half maximum of 67.8 cells. This correlation, which is excellent, is shown in the figure at the top right. The data from our billion-cell simulation of homogeneous, compressible turbulence, described in the first part of this article, supports this same model for F_{SGS} with an equally strong correlation to that shown in the figure here, even though in that flow the divergence of the filtered velocity field, with an *rms* Mach number of about 1/3, tends to dominate the term in Π_f on the right in the above relationship. For this data, as described earlier, two filters are used, with full widths at half maximum of 67.8 and 6.03 cells.

The relation for F_{SGS} given above can perhaps be used in building a k-ε model of subgrid-scale turbulence. In this case, a model for the subgrid-scale stress, τ_{ij}, that produces this relation for F_{SGS} is $\tau_{ij} = k\delta_{ij} + A\Pi_{ij}$, where k, the subgrid-scale turbulent kinetic energy, τ_{ii}, is approximated by $\lambda_f^2 \overline{\rho}|\partial_i \tilde{u}_j|^2$ and where $A = (\lambda_f^2 \overline{\rho}/2)(\det(\Pi)/|\Pi|^2)$ can be either positive or negative. This suggests a mechanism for incorporating the model into the momentum and total energy conservation laws of a numerical scheme such as PPM while maintaining the scheme's strict conservation form. The approximation of k, the subgrid-scale turbulent kinetic energy, involving velocity changes on the scale of the filter demands that, in proper LES style, the larger turbulent eddies are resolved on the grid, so that these velocity changes are meaningful. In the spirit of a k-ε model, we could replace $\lambda_f^2 \, \overline{\rho} |\, \partial_i \widetilde{u}_j \,|^2$ by k in the relation for F_{SGS} and use the resulting form for F_{SGS} as the time rate of change of k in a frame moving with the velocities $\widetilde{u}_j$, that is, with the resolved velocity field in the LES calculation. A further term, related to the ε of a k-ε model, representing the decay of k due to viscous dissipation on scales well below that of the grid would also have to be included. We would then have a dynamical partial differential equation for k to solve along with the conservation laws for mass, momentum, and total energy. Constructing such a subgrid-scale model of turbulence for use with the PPM gas dynamics scheme is a subject of future work.

4   Acknowledgements

We would like to acknowledge generous support for this work from the Department of Energy, through grants DE-FG02-87ER25035 and DE-FG02-94ER25207, contracts from Livermore and Los Alamos, and an ASCI VIEWS contract through NCSA; from the National Science Foundation, through its PACI program at NCSA, through a research infrastructure grant, CDA-950297, and also through a Grand Challenge Application Group award, ASC-9217394, through a subcontract from the University of Colorado; and from NASA, through a Grand Challenge team award, NCCS-5-151, through a subcontract from the University of Chicago. We would also like to acknowledge local support from the University of Minnesota's Minnesota Supercomputing Institute. This work was performed in part under the auspices of the U.S. DoE by the Lawrence Livermore National Laboratory under contract NO. W-7405-ENG-48.

References

1.   P. R. Woodward and P. Colella, "The Numerical Simulation of Two-Dimensional Fluid Flow with Strong Shocks," *J. Comput. Phys.* **54**, 115–173 (1984).
2.   P. Colella and P. R. Woodward, "The Piecewise-Parabolic Method (PPM) for Gas Dynamical Simulations," *J. Comput. Phys.* **54**, 174–201 (1984).
3.   P. R. Woodward, "Numerical Methods for Astrophysicists," in *Astrophysical Radiation Hydrodynamics*, K.-H.Winkler and M. L. Norman, eds., 245–326 (Reidel, 1986).
4.   B. K. Edgar, P. R. Woodward, S. E. Anderson, D. H. Porter and Wenlong Dai (1999). PPMLIB home page at *http://www.lcse.umn.edu/PPMlib*.

5. P. R. Woodward and S. E. Anderson, SPPM benchmark code, LCSE version (1995). Available at *http://www.lcse.umn.edu*.

6. D. H. Porter and P. R. Woodward, "High Resolution Simulations of Compressible Convection with the Piecewise-Parabolic Method (PPM)," *Astrophysical Journal Supplement*, **93**, 309–349 (1994).

7. *http://www.llnl.gov/casc/asciturb* — site for ASCI turbulence team at LLNL.

8. J. P. Boris, F. F. Grinstein, E. S. Oran, and R. J. Kolbe, "New Insights into Large Eddy Simulation," *Fluid Dyn. Res.* **19**, 19 (1992).

9. C. Fureby, and F. F. Grinstein, "Monotonically integrated large eddy simulation of free shear flows," *AIAA J.* **37**(5), 544 (1999).

10. G.-S. Karamanos and G. E. Karniadakis, "A Spectral Vanishing Viscosity Method for Large-Eddy Simulations," *J. Comput. Phys.* **163,** 22–50 (2000).

11. Sytine, I. V., Porter, D. H., Woodward, P. R., Hodson, S. W., and Winkler, K.-H., 2000. "Convergence Tests for Piecewise Parabolic Method and Navier–Stokes Solutions for Homogeneous Compressible Turbulence," *J. Comput. Phys.*, **158**, 225–238.

12. D. H. Porter, A. Pouquet and P. R. Woodward, "A Numerical Study of Supersonic Homogeneous Turbulence," *Theoretical and Computational Fluid Dynamics*, **4**, 13–49 (1992).

13. D. H. Porter, P. R. Woodward and A. Pouquet, "Inertial Range Structures in Decaying Turbulent Flows," *Physics of Fluids* **10**, 237–245 (1998).

14. M. Vetter and B. Sturtevant, "Experiments on the Richtmyer–Meshkov Instability of an Air/SF6 Interface," *Shock Waves*, **4**, 247–252 (1995).

15. R. H. Cohen, W. P. Dannevik, A. M. Dimits, D. E. Eliason, A. A. Mirin, Y. K. Zhou, D. H. Porter, and P. R. Woodward, "Three-Dimensional Simulation of a Richtmyer–Meshkov Instability with a Two-Scale Initial Perturbation," in preparation.

16. D. H. Porter, P. R. Woodward, and M. L. Jacobs, "Convection in Slab and Spheroidal Geometries," Proc. 14th International Florida Workshop in Nonlinear Astronomy and Physics: Astrophysical Turbulence and Convection, Univ. of Florida, Feb., 1999; in *Annals of the New York Academy of Sciences* **898**, 1–20 (2000). Available at *http://www.lcse.umn.edu/convsph* .

17. *http://www.lcse.umn.edu/Movies* — LCSE web site to visit in order to view movies of the simulations discussed in this article. See also other links at *http://www.lcse.umn.edu*, including subdirectories *research/lanlrun*, *research/nsf*, *research/nsf/penconv96dec.html*, *research/RedGiant*. A PDF version of this article, with high-quality color illustrations, can be found at this Web site, along with animations from the computer simulations discussed here.

18. D. H. Porter and P. R. Woodward, "3-D Simulations of Turbulent Compressible Convection," *Astrophysical Journal Supplement*, **127**, 159–187 (2000).

INSTABILITIES AND FILAMENTATION
OF DISPERSIVE ALFVÉN WAVES

D. LAVEDER, T. PASSOT AND P. L. SULEM

CNRS UMR 6529, Observatoire de la Côte d'Azur
BP 4229, 06304 Nice Cedex 4, France
E-mail:sulem@obs-nice.fr

Direct numerical simulations of the three-dimensional Hall–MHD equations and
of a long-wavelength asymptotic model are used to study the instabilities and the
nonlinear dynamics of a circularly polarized Alfvén wave subject to a weak random
noise. The evolution is shown to be strongly sensitive to the spectral extension of
the initial noise, due to the presence of competing instabilities. The formation of
magnetic filaments is usually observed when only large-scale modulational perturbations are permitted, while a more turbulent picture is obtained when small-scale
unstable modes are initially excited. A filamentary dynamics nevertheless develops
in the presence of a broad initial spectrum in the case of a right-hand polarized
pump of long wavelength.

1 Introduction

The magnetohydrodynamic (MHD) description of magnetized plasmas con-
centrates on length scales that are much longer than the ion inertial length
c/ω_{pi}, defined as the ratio of the light velocity to the ion plasma frequency,
and time scales much larger than the inverse ion gyromagnetic frequency Ω_i^{-1}.
When dealing with phenomena whose characteristic scales approach these limits, the Hall term that originates from the ion inertia becomes relevant in the
generalized Ohm's law. In the absence of significant dissipation, the dynamics
is then governed by the Hall–MHD equations[1]

$$\partial_t \rho + \nabla \cdot (\rho \mathbf{u}) = 0 \tag{1}$$

$$\rho(\partial_t \mathbf{u} + \mathbf{u} \cdot \nabla \mathbf{u}) = -\frac{\beta}{\gamma}\nabla \rho^\gamma + (\nabla \times \mathbf{b}) \times \mathbf{b} \tag{2}$$

$$\partial_t \mathbf{b} - \nabla \times (\mathbf{u} \times \mathbf{b}) = -\frac{1}{R_i}\nabla \times \left(\frac{1}{\rho}(\nabla \times \mathbf{b}) \times \mathbf{b}\right) \tag{3}$$

$$\nabla \cdot \mathbf{b} = 0. \tag{4}$$

As usual, ρ is the density of the plasma, $\mathbf{u}$ its velocity and $\mathbf{b}$ the magnetic
field. The equations are written in a nondimensional form, taking as unity
the Alfvén speed $c_A = \mathcal{B}_0/\sqrt{4\pi\rho_0}$, where $\mathcal{B}_0$ is the magnitude of the ambient
magnetic field and ρ_0 the mean density of the plasma. Furthermore, $\beta = c_s^2/c_A^2$, where c_s is the sound velocity, γ is the polytropic gas constant and

$R_i = \Omega_i L/c_A$, where L is a reference length, denotes the nondimensional ion–cyclotron frequency.

In the presence of an ambient magnetic field taken in the x direction, the Hall–MHD equations admit special solutions, named after Ferraro,[2] in the form of finite-amplitude circularly-polarized Alfvén waves $b_y - i\sigma b_y = B_0 e^{i(kx-\omega t)}$ propagating along the magnetic field with, for forward propagation, a dispersion relation $\omega = \frac{\sigma k^2}{2R_i} + k\sqrt{1 + \left(\frac{k}{2R_i}\right)^2}$ where $\sigma = +1$ or -1, depending on right-hand or left-hand polarization. An important issue concerns the instabilities that can affect such pump waves and the forthcoming nonlinear dynamics that can lead to small-scale formation and heating of the plasma. For purely longitudinal perturbations, the problem becomes one-dimensional and was extensively studied. Depending on the situation, solitonic structures, envelope solitons or a turbulent cascade are obtained.[3]

Another special case corresponds to the effect of transverse perturbations. For small amplitude waves, the problem is amenable to a multiple-scale analysis leading to a two-dimensional nonlinear Schrödinger (NLS) equation[4] for the pump envelope that, in some instances, predicts wave collapse and formation of intense magnetic filaments (Alfvén wave filamentation).[5-7]

When scale separation does not hold, as a consequence of a finite-amplitude wave or of instabilities at the scale of the pump, the problem is to be addressed by numerical simulations. It nevertheless simplifies when the Alfvén wave is considered in the long-wavelength limit. By means of a reductive perturbative expansion, the Hall–MHD equations then lead to long-wave asymptotic equations that can be viewed as a three-dimensional generalization[8,9] of the derivative nonlinear Schrödinger (DNLS) equation[10-14], including the coupling to the magnetosonic waves averaged along the direction of propagation. These equations provide in particular a description of transverse collapse in the case of finite amplitude Alfvén waves where filamentation is replaced by the formation of "magnetic pancakes" with moderate amplification of the wave amplitude but development of strong gradients.[15]

Simulations are presented here for various regimes of parameters. The algorithms used for the three-dimensional Hall–MHD and DNLS equations are based on a Fourier spectral method and a third-order Runge–Kutta time stepping discussed in the appendix. In all the simulations we used $c_A/\Omega_i = c/\omega_{pi}$ as the reference scale L, which implies $R_i = 1$. A main observation is the sensitivity of the nonlinear dynamics not only to the characteristics of the pump wave but also, because of the presence of various competing instabilities, to the spectral extension of the initial perturbating noise.

18

2 The special case of longitudinal dynamics

A first illustration of the multiplicity of regimes developed by dispersive Alfvén waves propagating along the ambient field is already obtained in one space dimension. In the plane (k, β), the ranges of existence of the various linear instabilities affecting a small-amplitude circularly polarized pump of wavenumber k appear to be delimited by the resonances associated with the equality

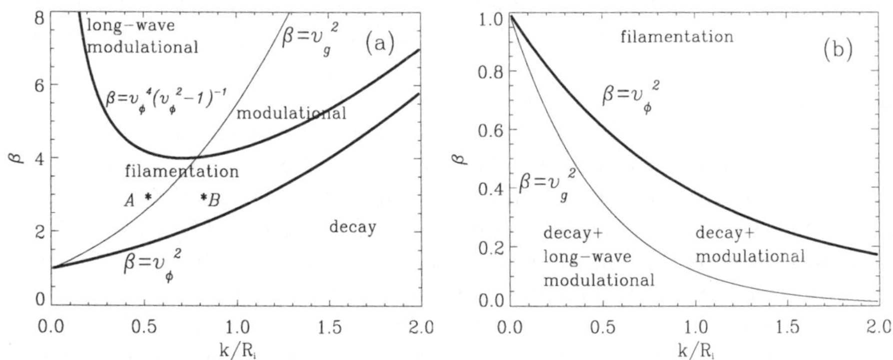

Figure 1. Instability regions in the plane (β, k) for a small-amplitude Alfvén wave with right-hand (a) or left-hand (b) polarization. Thick lines delimit the domains of the filamentation instability. Points A and B correspond to typical conditions in forthcoming numerical simulations.

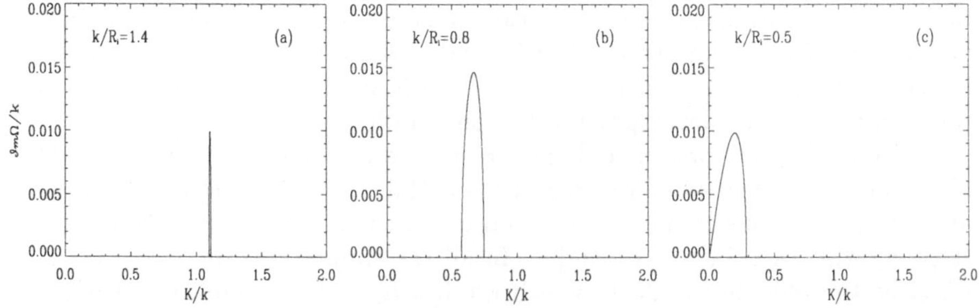

Figure 2. Growth rate versus perturbation wavenumber (both normalized by the pump wavenumber k) for examples of decay (a), modulational (b) and long-wavelength modulational (c) instabilities, for a right-hand polarized wave when $\beta = 2.7$.

of the sound speed (given by $\sqrt{\beta}$ when the Alfvén velocity is taken as unity) with the group or phase velocities v_g or v_ϕ of the Alfvén wave[16] (Fig. 1). Note that different instabilities can coexist in the case of a left-hand polarized wave. As exemplified in Fig. 2(a), decay instability affects wavenumbers larger than the pump wavenumber k, while the instability is said modulational in the Alfvén wave literature[17] when the unstable wavenumbers are smaller than k (Fig. 2(b)). The rescaling by the pump wavenumber in Fig. 2 is suggested by the fact that for a wave perturbation of the form $e^{i[(k\pm K)x-(\omega\pm\Omega)t]}$, the dispersion relation implies that Ω/k depends on K/k and k/R_i. In the special case of a "long-wavelength" modulational instability that extends down to the zero wavenumber (Fig. 2(c)), the forthcoming nonlinear dynamics is amenable to a multiple-scale analysis that leads to a nonlinear Schrödinger (NLS) equation

$$i\partial_\tau B + \frac{\omega''}{2}\partial_{\xi\xi}B + \frac{kv_g}{4(\beta - v_g^2)}|B|^2 B = 0. \tag{5}$$

This equation describes in terms of the stretched variables $\xi = \epsilon(x - v_g t)$ and $\tau = \epsilon^2 t$, the evolution of the Alfvén wave amplitude defined, up to subdominant harmonics, by $b_y - i\sigma b_z = \epsilon B e^{i(kx-\omega t)}$ with $\sigma = -1$ for left-hand polarization and $\sigma = +1$ for right-hand polarization. In one space dimension, the cubic NLS equation is integrable and predicts the formation of envelope solitons. In contrast, when the instability takes place at smaller scales (Figs. 2(b) and 2(c)), the coupling to the magnetosonic waves plays an important role, leading to strong nonlinear phenomena with formation of density shocks.

When the pump wavenumber k is decreased to order ϵ^2, while the wave amplitude is kept of order ϵ, the dispersion becomes comparable to the nonlinearity and the dynamics is no longer governed by an envelope equation. It is nevertheless amenable to a reductive perturbative expansion which, for $\beta \neq 1$, selects the Alfvén waves. To leading order, they are governed by the previously mentioned DNLS equation for the transverse magnetic field $b_y + ib_z = \epsilon b$

$$\partial_\tau b + \frac{1}{4(1-\beta)}\partial_\xi \left((|b|^2 - \langle|b|^2\rangle)b \right) + \frac{i}{2R_i}\partial_{\xi\xi}b = 0, \tag{6}$$

where the brackets indicate averaging along the direction of propagation. The stretched variables are defined as $\xi = \epsilon^2(x - t)$ and $\tau = \epsilon^4 t$. Like many other long-wavelength equations (Korteweg–de Vries, Benjamin–Ono, etc.), the DNLS equation is a soliton equation integrable by inverse scattering.[18] It involves a complex field because of the degeneracy associated with the equality in the dispersionless limit of the phase velocity of the Alfvén and

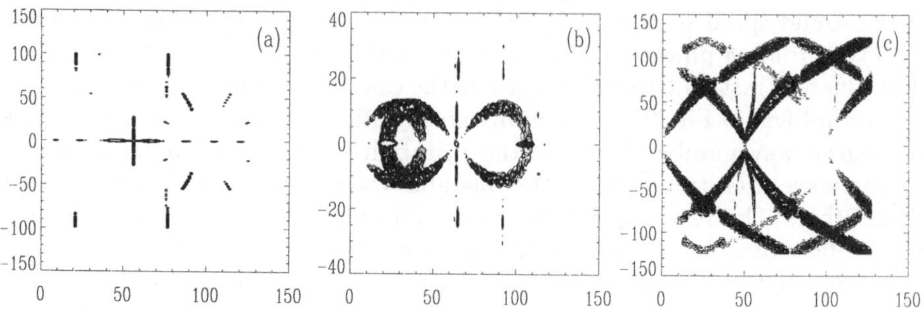

Figure 3. Spectral instability range as delimited by b_y Fourier modes contours (similar picture with b_z) obtained by two-dimensional numerical simulations of Hall–MHD equations, for a small-amplitude right-hand polarized Alfvén wave with $k = 0.5$ (a) or $k = 0.8$ (b) both for $\beta = 2.7$, and for a left-hand polarized wave with $k = 1$ when $\beta = 1.5$. The (positive) longitudinal and transverse wavenumbers are measured in units of $\Delta k_\parallel = 1/112$ (a), $1/80$ (b), $1/50$ (c) and $\Delta k_\perp = 1/72$ (a), $1/96$ (b), $1/28$ (c).

fast ($\beta < 1$) or slow ($\beta > 1$) magnetosonic waves in the case of propagation along the ambient field. It is convenient to introduce a parameter ν, measuring the relative magnitude of the dispersion compared to the nonlinearity. For this purpose, denoting by b_0 and k^{-1} the initial typical amplitude and (longitudinal) wavelength of the transverse magnetic field, we rescale b by b_0, the longitudinal coordinate by k^{-1} and the time by $(kb_0^2)^{-1}$. In Eq. (6) the coefficient $\frac{1}{2R_i}$ is then replaced by $\nu = \frac{k}{2R_ib_0^2}$. In the limit of large ν (very small amplitude wave), the dispersion is dominant and a multiple-scale analysis performed on the DNLS equation reproduces the NLS equation (5) where the coefficients are taken in the limit $k \to 0$. The DNLS equation that does not retain counterpropagating waves cannot however describe the decay instability (that survives in the long-wavelength limit) and is thus restricted to plasmas with $\beta > 1$. In this case, quantitative comparisons with the direct numerical simulations[16] show a satisfactory agreement on times that scale like ϵ^{-4}, which validates the asymptotics.

3 Multidimensional linear instabilities

The longitudinal instabilities displayed in Figs. 1 and 2 have multidimensional extensions and coexist with transverse and oblique instabilities, as seen in Fig. 3 for a right-hand polarized wave in typical conditions corresponding to points A ($k = 0.5$, $\beta = 2.7$) and B ($k = 0.8$, $\beta = 2.7$) of Fig. 1(a) and for left-hand polarization with $k = 1$ and $\beta = 1.5$. The instability ranges are obtained by short-time integrations of the Hall–MHD equations, restricted to two space dimensions because of the symmetry of the problem with respect to the direction of the transverse component of the perturbation wavevector. The nonlinear dynamics is in contrast sensitive to the presence of a third dimension.

As the pump wavenumber is reduced, the dynamics considerably slows down and it is convenient in this regime to use a three-dimensional extension of the DNLS equation discussed in Section 2. This long-wavelength limit involves a scaling of the transverse variables $\eta = \epsilon^3 y$, $\zeta = \epsilon^3 z$ and must include the coupling to mean fields (denoted by bars or brackets) resulting from the averaging of magnetosonic waves along the direction of propagation[8,9]

$$\partial_\tau b + \partial_\xi \left(\frac{1}{2} bP + (\bar{u}_x + \frac{1}{2}\bar{b}_x)b \right) - \frac{1}{2}\partial_\perp P + \frac{i}{2R_i}\partial_{\xi\xi} b = 0 \tag{7}$$

$$\partial_\tau \bar{u}_x = \frac{1}{2}\left(\partial_\perp^* \langle bP \rangle + \partial_\perp \langle b^* P \rangle \right) \tag{8}$$

$$\partial_\xi \tilde{b}_x + \frac{1}{2}\left(\partial_\perp^* b + \partial_\perp b^* \right) = 0. \tag{9}$$

In the above 3D-DNLS system, the notation $\partial_\perp = \partial_\eta + i\partial_\zeta$ has been introduced. Like the mean longitudinal velocity $\bar{u}_x$, the induced longitudinal magnetic field is rescaled by a factor ϵ and is separated in mean and fluctuating parts in the form $\bar{b}_x + \tilde{b}_x$. Furthermore, $\bar{b}_x = -\frac{1}{1+\beta}\left(\frac{\langle |b|^2 \rangle}{2} - E_M \right)$ where the constant E_M, defined as the average over the whole domain of the magnetic energy density $\frac{|b|^2}{2}$, is retained to ensure that $\bar{b}_x$ is zero in one space dimension. The fluctuations of magnetic pressure are given by $P = \frac{1}{2(1-\beta)}\left(2\tilde{b}_x + |b|^2 - \langle |b|^2 \rangle \right)$. By means of the same change of variables as in one dimension, together with the rescaling of the transverse coordinates by $(b_0 k)^{-1}$ and of the fields $\bar{u}_x$, $\bar{b}_x$, $\tilde{b}_x$ by b_0^2, the coefficient $\frac{1}{2R_i}$ is replaced by the parameter ν in Eq. (7). Numerical simulations of Eqs. (7)–(9) show that the range of the unstable modes for a left-hand polarized pump in the long-wavelength limit remains qualitatively similar to that of Fig. 3(c), with the persistence of significant oblique instabilities. In contrast, the spectral instability range strongly simplifies in the case of a right-hand pump and concentrates to large-scale modes in the purely transverse and longitudinal directions. In this case, for both large ($\nu = 50$) and moderate ($\nu = 0.5$) dispersion regimes, the growth rate in the transverse direction is dominant by about one order of magnitude when $\beta = 3$. As checked on direct numerical simulations of the Hall–MHD equations with decreasing pump wavenumber, the other instabilities are either suppressed or moved to scales that are too small to be retained by the DNLS asymptotics.

4   Transverse dynamics and filamentation

As already mentioned, special attention was devoted to the transverse instability that affects low frequency modes and is amenable to a multiple-scale analysis leading to a two-dimensional NLS equation[6]

$$i\partial_\tau B + \alpha\Delta_\perp B - kv_g\left(\frac{1}{v_g^2} - \frac{k^4}{4(\beta+1)\omega^4}\right)|B|^2 B = 0. \tag{10}$$

Here the Laplacian is taken relatively to the stretched transverse variables $Y = \epsilon y$ and $Z = \epsilon z$. It follows that a right-hand polarized pump wave is unstable relatively to a long-wavelength transverse modulation (filamentation instability) when $v_\phi^2 < \beta < \frac{v_\phi^4}{v_\phi^2-1}$, while a left-hand polarized pump wave is unstable when $\beta > v_\phi^2$ (Fig. 1). In one dimension, the nonlinear saturation of the instability results in the formation of solitonic structures, while much more violent effects take place in two dimensions where the wave focuses transversally and collapses with a local blowup of the field B, associated with a breakdown of the multiple-scale asymptotics. In the primitive variables, this corresponds to the formation of intense magnetic filaments parallel to the ambient field. The question however arises of the relevance of the filamentation phenomenon in a context where the dynamics is not restricted to purely transverse instabilities. As seen in Fig. 3, the spectral ranges of the instabilities that develop for a pump of moderate wavenumber are rather extended and one may expect that the dynamics resulting from a competition of several instabilities will be sensitive to the characteristics of the initial noise.

As already noticed, the situation strongly simplifies in the DNLS description of a right-hand polarized pump, due to the absence of oblique instabilities. Numerical simulations of Eqs. (7)–(9) in the strongly dispersive regime ($\nu = 50$) with a broad-spectrum initial noise show the formation of magnetic filaments.[19] In order to resolve the phenomenon of wave filamentation more accurately, in a context where only the transverse dynamics turns out to be relevant, we chose to present here simulations in a periodic box whose longitudinal extension equals the wavelength of the pump.[15] Figure 4(a) displays the filamentation that occurs for $\nu = 50$ when $\beta = 3$. The transverse cut presented in Fig. 4(b) shows that at the end of the simulation, the intensity $|b|^2$ has been locally amplified by two orders of magnitude. When the strength of the dispersion is reduced ($\nu = 0.5$), the transverse dynamics is no longer governed by the two-dimensional NLS equation. In this case, the amplification

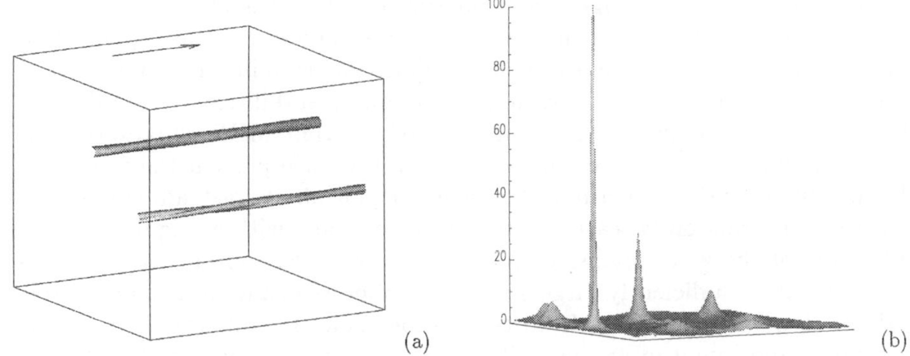

Figure 4. Magnetic filaments where $|b|^2$ exceeds its initial value by a factor 10 (a) and snapshot of $|b|^2$ in a plane transverse to the ambient field (b) in numerical simulations of the 3D-DNLS system with $\beta = 3$ and $\nu = 50$ at resolution 32×256^2. The arrow refers to the direction of the ambient magnetic field.

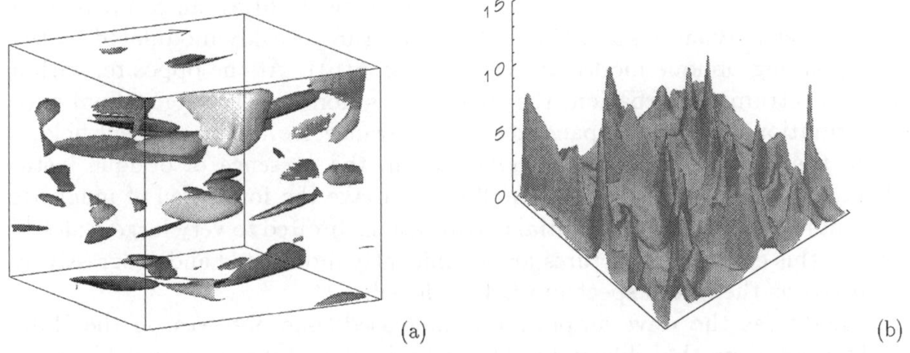

Figure 5. "Magnetic pancakes" where $|b|^2$ exceeds its initial value by a factor 4 (a) and snapshot of $|b|^2$ in a plane transverse to the ambient field (b) in numerical simulations of the 3D-DNLS system with $\beta = 3$ and $\nu = 0.5$ at resolution 32×128^2.

of the transverse magnetic field is only moderate and its intensity displays significant variations in the longitudinal direction, leading to the formation of magnetic structures with a pancake shape (Fig. 5(a)). Strong gradients develop in planes transverse to the ambient field (Fig. 5(b)) and the question arises of the existence of a finite-time gradient singularity.

Still in the case of a right-hand polarized pump, it is of interest to consider the effect of increasing the wavenumber to moderate values (point A of Fig. 1(a)) for which the longitudinal modulational instability remains at large scale with nevertheless the presence of oblique instabilities at small scales (Fig. 3(a)). Direct numerical simulations of the Hall–MHD equations show that if small-scale modes are initially excited, they will prevent the formation of coherent magnetic structures (filaments or pancakes). Instead, small-scale structures are randomly scattered within the domain, with no significant amplification of the wave intensity. In contrast, when the initial perturbations are confined at sufficiently large scales, the oblique instabilities play no role and the dynamics, resulting from the competition of comparable large-scale transverse and longitudinal instabilities, leads to the formation of elongated structures where both the modulation associated with the longitudinal instability and the effect of the pump harmonics are visible (Fig. 6(a)).

When increasing further the pump wavenumber, the longitudinal instability no longer affects the largest scales but rather those comparable to the pump wavelength (point B of Fig. 1(a)). Two classes of perturbations are then considered. For an initial noise at scales large compared to that of the longitudinal instability, only large-scale oblique instabilities can compete with the transverse dynamics and filamentary structures weakly modulated by the corresponding oblique modes are visible (Fig. 6(b)). At the opposite, with a broad-spectrum perturbation, the dynamics is dominantly longitudinal with the formation of magnetic pancakes mostly transverse to the ambient field.

In the context of left-hand polarization, the presence of oblique instabilities extending from large to small scales make the formation of magnetic filaments possible only with initial perturbations limited to very large scales.[19] Even in this case, the structures are significantly modulated and progressively destroyed as the initial spectrum is broadened.

Finally, as the wave amplitude is increased, one observes on the Hall–MHD simulations that, like in the DNLS limit, the intense magnetic filaments are replaced by magnetic pancakes of moderate amplitude.

5 Concluding remarks

The instabilities and the further nonlinear evolution of a circularly polarized Alfvén wave are analyzed in the context of Hall–MHD that retains dispersive effects due to ion inertia. The strong sensitivity of the dynamics to the spectral extension of the initial noise is in particular pointed out. Furthermore, the transverse collapse of a small-amplitude wave, leading to the formation of intense magnetic filaments parallel to the ambient field, is shown to be especially relevant for long-wavelength pumps with right-hand polarization, even in the presence of broad-spectrum perturbations that in other regimes may induce a more turbulent regime. Wave filamentation provides a mechanism for small scale formation that permits acceleration and heating of the plasma in a regime where no small-scale instabilities are present, and thus without requiring large-amplitude waves or inhomogeneity of the medium. Future developments should include the influence of finite ion Larmor radius [20,21] and the role of kinetic effects that are believed to be relevant when the parameter β exceeds unity. In one space dimension, a simple description of the latter effects in the long-wave limit is provided by additional nonlocal terms in the DNLS equation.[22-24]

Appendix: Numerical schemes

Both the three-dimensional Hall–MHD and DNLS equations are integrated using a pseudospectral method based on Fourier mode expansion. The dynamics being strongly sensitive to the linear instabilities that affect the system, a special attention is to be paid to the choice of the spatial discretization, in order to retain the most unstable modes. This may require the use of a computational box with different sizes in the longitudinal and transverse directions.

Concerning the time stepping, the Hall–MHD equations are solved with

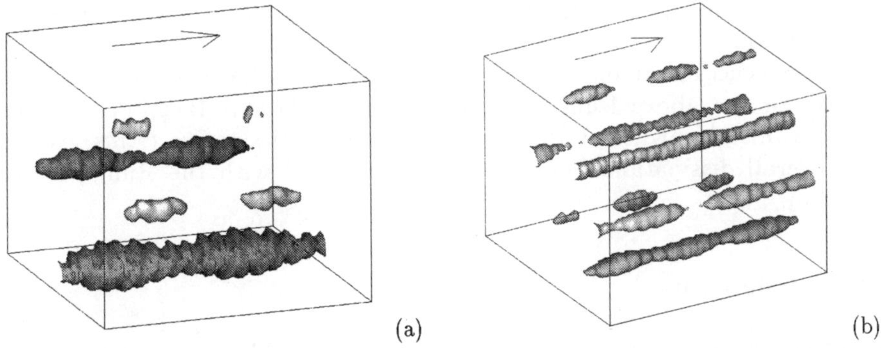

Figure 6. Regions where the transverse magnetic field intensity exceeds its initial value by a factor 4.5 in the conditions of point A (a) and by a factor 3.5 in the conditions of point B (b), resulting from the instability of an Alfvén wave of initial amplitude 0.1, in numerical simulation of Hall–MHD equations at resolution $128 \times (64)^2$.

a low-storage Runge–Kutta scheme.[25] For the DNLS system, symbolically written

$$\frac{dU}{dt} = N(U) + LU,$$

the same scheme is used for the nonlinearities $N(U)$, while the Hall term LU that in this case is linear and purely longitudinal, is treated exactly, using an exponential scheme. The transition from t_n to $t_{n+1} = t_n + \delta t$ is then given by

$$V_1 = e^{\frac{1}{3}\delta t L}\left(U_n + \tfrac{1}{3}H_1\right) \qquad H_1 = \delta t N(U_n)$$

$$V_2 = e^{\frac{5}{12}\delta t L}\left(V_1 + \tfrac{15}{16}H_2\right) \qquad H_2 = \delta t N(V_1) - \frac{5}{9}e^{\frac{1}{3}\delta t L}H_1$$

$$U_{n+1} = e^{\frac{1}{4}\delta t L}\left(V_2 + \tfrac{8}{15}H_3\right) \qquad H_3 = \delta t N(V_2) - \frac{153}{128}e^{\frac{5}{12}\delta t L}H_2.$$

This numerical scheme is slightly dissipative, of an amount consistent with the $(\delta t)^3$-accuracy, which contributes to the numerical stability.

For the Hall–MHD equations, the maximal time step δt allowing numerical stability varies like N^{-2} in terms of the number N of collocation points in each direction. The integrations are also very time consuming due to the slow growth rate of the modulational instabilities, while the time step (typically $\delta t = 1.25 \cdot 10^{-3}$ for $N = 128$ in the conditions of our simulations that include a spherical truncation of the spectral domain) is constrained by stability conditions. As a consequence, the three-dimensional simulations were limited to this resolution.

We also checked on one-dimensional integrations, that with the time step retained for the above Runge–Kutta scheme, an Adams–Bashforth scheme would be unstable without the inclusion of extra dissipative terms leading to an overall dissipation larger than that obtained with the Runge–Kutta algorithm.

Acknowledgments

Computations were performed on the CRAY C94, CRAY C98 and NEC machines of the Institut du Développement et des Ressources en Informatique Scientifique and on the facilities of the Program "Simulations Interactives et Visualisation en Astronomie et Mécanique" at the Observatoire de la Côte d'Azur. This work was supported in part by the "Programme National Soleil Terre" of the CNRS and by NSF Grant No. PHY94-07194.

References

1. J. D. Huba, *Phys. Plasmas* **2**, 2504 (1995).
2. V. C. A. Ferraro, *Proc. R. Soc. London* Ser. A **233**, 310 (1955).
3. S. R. Spangler, in *Nonlinear Waves and Chaos in Space Plasmas*, 171, T. Hada and H. Matsumoto, eds., (Terra Scientific, Tokyo, 1997).
4. C. Sulem and P. L. Sulem, *The Nonlinear Schrödinger Equation*, Applied Mathematical Sciences **139** (Springer-Verlag, 1999).
5. P. K. Shukla and L. Stenflo, *Space Sci.* **155**, 145 (1989).
6. S. Champeaux, T. Passot and P. L. Sulem, *J. Plasma Phys.* **58**, 665 (1997).
7. S. Champeaux, A. Gazol, T. Passot and P. L. Sulem, in *Nonlinear MHD Waves and Turbulence*, T. Passot and P. L. Sulem, eds., Lecture Notes in Physics **536**, 54 (Springer-Verlag, 1999).
8. T. Passot and P. L. Sulem, *Phys. Rev. E* **48**, 2966 (1993).
9. A. Gazol, T. Passot and P. L. Sulem, *Phys. Plasmas* **6**, 3114 (1999).
10. A. Rogister, *Phys. Fluids* **14**, 2733 (1971).
11. K. Mio, T. Ogino, K. Minami and S. Takeda, *Phys. Soc. Japan* **41**, 265 (1975).
12. E. Mjølhus, *J. Plasma Phys.* **16**, 321 (1976).
13. C. F. Kennel, B. Buti, T. Hada and R. Pellat, *Phys. Fluids* **31**, 1949 (1988).
14. E. Mjølhus and T. Hada in *Nonlinear Waves and Chaos in Space Plasmas*, 121, T. Hada and H. Matsumoto, eds. (Terra Scientific, Tokyo, 1997).
15. D. Laveder, T. Passot and P. L. Sulem, Transverse collapse of low-frequency Alfvén waves, to appear in *Physica D*.
16. S. Champeaux, D. Laveder, T. Passot and P. L. Sulem, *Nonlin. Proc. Geophys.* **6**, 169 (1999).
17. H. K. Wong and M. L. Goldstein, *J. Geophys. Res.* **91**, 5617 (1986).
18. D. J. Kaup and A. C. Newell, *J. Math. Phys.* **19**, 798 (1978).
19. D. Laveder, T. Passot and P. L. Sulem, Three-dimensional dynamics of Alfvén waves, preprint.
20. K. V. Roberts and J. B. Taylor, *Phys. Rev. Lett.* **8**, 197 (1962).
21. V. A. Marchenko, R. D. Denton, M. K. Hudson, *Phys. Plasmas* **3**, 3861 (1996).
22. E. Mjølhus and J. Wyller, *J. Plasma Phys.* **40**, 299 (1988).
23. S. R. Spangler, *Phys. Fluids B* **2**, 407 (1990).
24. M. V. Medvedev and P. H. Diamond, *Phys. Plasmas* **3**, 863 (1995).
25. C. Canuto, Y. Hussaini, A. Quateroni and T. A. Zang, *Spectral Methods in Fluid Dynamics* (Springer Verlag, 1988).

ASTROPHYSICAL JETS

A. C. RAGA[1], H. SOBRAL[2], M. VILLAGRAN-MUNIZ[2], R. NAVARRO-GONZALEZ[3], S. CURIEL[1], L. F. RODRIGUEZ[4], J. CANTO[1]

[1] *Instituto de Astronomía, UNAM, Apdo. Postal 70-264, 04510 D. F., México*
[2] *Centro de Instrumentos, UNAM, Apdo. Postal 70-186, 04510 D. F., México*
[3] *Instituto de Ciencias Nucleares, UNAM, Apdo. Postal 70-543, 04510 D. F., México*
[4] *Instituto de Astronomía, UNAM, Apdo. Postal 74-3 (Xangari), 58089 Morelia, Michoacán, México*

1 Introduction

Astrophysical jets span an enormous range of characteristic sizes, with lengths ranging from $\sim 10^{15}$ cm (e.g., the "microjets" from T Tauri stars) to $\sim 10^{25}$ cm (for powerful jets from radio galaxies). These sizes are of course in all cases enormous relative to the sizes of "aeronautical" or laboratory jets, which range from ~ 1 cm to $\sim 10^{4}$ cm.

However, the differences between astrophysical and laboratory jets are not as large as the above discussion appears to suggest. While some extragalactic jets appear to have Mach numbers of order unity, most astrophysical jets are hypersonic, with a $M \sim 5\text{--}100$ Mach number range. As there are published results of laboratory jets with Mach numbers of up to ~ 20 (see, e.g., Harvey and Hunter[1]), it is clear that the experiments do get into the range of interest for astrophysical flows.

Interestingly, it is hard to estimate the Reynolds number of many astrophysical jets. The mean free path of the ions and electrons in an extragalactic jet is larger than the diameter of the jet beam, resulting in a low Reynolds number for the flow, and calling into question whether or not a gasdynamic description of this kind of flow is indeed appropriate at all. However, the gyroradius of the charged electrons is much smaller than the jet beam diameter, and magnetic field–particle interactions probably provide a large effective cross section. It is believed (though not proven in detail) that this kind of process does result in a fluid behavior for the jet flow, and that the effective Reynolds number of the flow is high. For the case of the much lower energy (and many times mostly neutral) jets ejected by stars, there is no doubt that the Reynolds number is indeed very high, with values in excess of 10^{5}.

Therefore, it is usually conjectured that astrophysical jets are in the "high Reynolds number regime" (Re> 10^5), in which the properties of the flow are independent of the actual value of Re. This high Reynolds number regime is easily reached in laboratory experiments of supersonic jets.

From the above discussion, we see that both the Mach and Reynolds numbers of astrophysical jets can be reached with laboratory experiments. There are, however, some interesting differences between astrophysical and "experimental" jets.

Both extragalactic and stellar jets show coherent shock structures that travel away from the jet source. These structures are produced either as a result of an intrinsic variability of the ejection, or as instabilites in the jet beam in the region close to the source. If one is to find such type of behavior in a laboratory jet, one needs to have a jet production mechanism that can vary over very short timescales and a very fast measuring system (since travelling shock waves will be advected down the jet beam in milliseconds). Such experiments are indeed possible and will be discussed below.

Finally, another important difference between astrophysical and laboratory jets lies in their radiative properties. While laboratory jets are adiabatic (which in the astrophysical vocabulary actually means "nonradiative"), having radiative energy losses which are small compared to the total energy of the jet, astrophysical jets range from the adiabatic regime (for extragalactic jets) to the highly radiative regime (for jets from young stars). Even though the effects of a radiative energy loss can be simulated in laboratory flows by introducing liquid droplets with an appropriate vaporization latent energy, we are not aware of any published results in which this kind of technique has been used for supersonic jet flows.

In this paper, we describe the general characteristics of a new, adaptive grid gasdynamic code (section 2). We then present a comparison of "starting jet" experiments with numerical simulations carried out with our code (section 3). Finally, we discuss the observational properties of a hypersonic jet produced by a recently formed star and present numerical simulations of this astrophysical object (section 4).

2 The "yguazú-a" code

We have developed a new code that integrates the gasdynamic equations, together with a system of rate equations for chemical/atomic/ionic species, in either two (cartesian or cylindrical) or three dimensions. The gasdynamic equations are integrated with a second order adaptation of the "flux vector splitting" method of Van Leer[2], and the microphysical rate terms are inte-

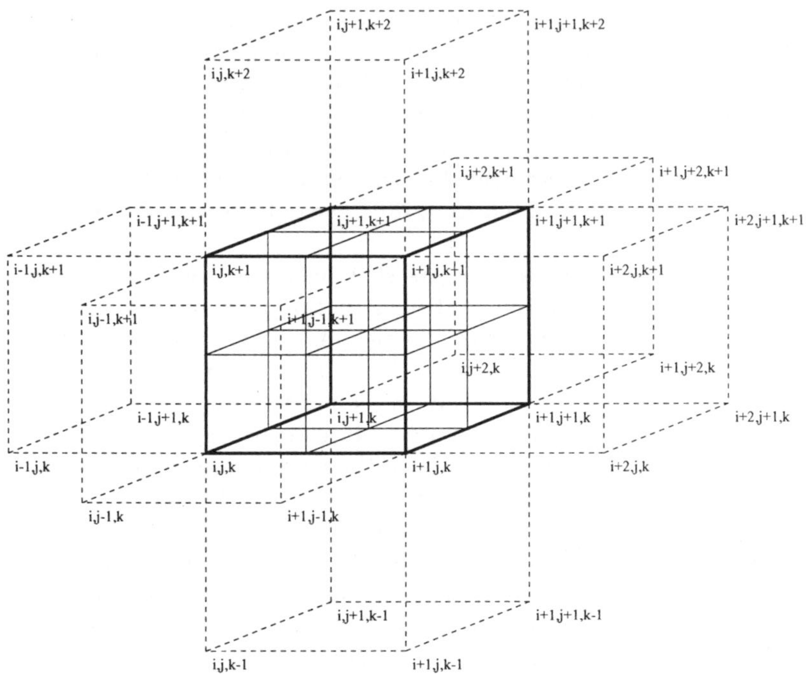

Figure 1. Schematic diagram showing a cube in grid g (with vertex i, j, k closest to the origin of the coordinate system), and the contiguous cubes of grid g. The thin, solid lines show the geometry of the points to be generated on refinement to grid $g + 1$. Eight of the grid $g + 1$ points coincide with the vertices of the grid g cube. Twelve of the grid $g + 1$ points lie along the edges of the grid g cube (e.g., point $i + 1/2, j, k$) and are assigned flow variables corresponding to the straight average of the primitive variables at the contiguous vertices. Six grid $g + 1$ points lie on the center of each of the faces of the cube and are assigned values corresponding to the average of the four vertices of the cube face. Finally, there is one grid $g + 1$ point at the center of the cube, to which we assign the average of the 8 vertices of the grid g cube.

grated with a semiimplicit method, which is described in detail by Raga *et al.*[3] It is also possible to simultaneously solve a radiative transfer problem (of the diffuse and direct radiation), with the method described by Raga *et al.*[4]

The most interesting feature of the yguazú-a code is its adaptive grid. The gasdynamic and rate equations are integrated on an adaptive, binary, hierarchical computational grid. The characteristics of the 3D grid are described in the following.

- Base grid: There are two grids, $g = 1$ and $g = 2$, which are defined over the whole computational domain. The grid spacings along the three coordinate axes of grid $g = 2$ are a factor of 2 smaller than the corresponding spacings of grid $g = 1$.

- Timestep: the timestep is chosen with the appropriate Courant condition, and the successive grids are marched forward in time with timesteps which differ by factors of 2.

- Refinements on roundoff error: As an estimate of the roundoff error, we use the difference between the values at a given spatial position obtained in two successive grids g and $g - 1$. If at a given time the integrated flow variables obtained for any of the vertices of a cube (of consecutive points) in the g grid differ from the results obtained in the $g - 1$ grid (for the points defined on both grids) by a fractional amount greater than a fixed lower limit ϵ, the vertices of the cube are copied over to a higher resolution grid $g + 1$ (unless this region of the computational domain is already defined in the $g + 1$ grid). The missing points in the region of the $g + 1$ grid contained within the cube are created through linear interpolations between the values at the vertices of the g grid cube (see Fig. 1).

- Refinement on proximity of higher resolution grid: If any of the vertices of a cube in grid g or any of the vertices of the adjacent cubes (see Fig. 1) is defined in grid $g + 2$, the cube is then copied over and interpolated into grid $g + 1$ (unless the region of the domain is already defined in $g + 1$).

- Other refinement criteria: It is straightforward to introduce other refinement criteria. For example, one can limit some of the grid refinements to a chosen spatial region, or to a given Lagrangian region of the fluid (which can be labeled with a passive scalar). Also, one can refine on the value of any of the flow variables, or on their gradients.

- Derefinements: When the refinement criteria are not met, the corresponding grid points are deleted.

- Grid boundaries: In carrying out each timestep in grid g, it is necessary to use the values of the neighboring points in the grid. If these points do not exist, the values for the appropriate position are computed by linearly interpolating the primitive variables in both space and time between the appropriate points of grid $g - 1$.

With the points discussed above, one can construct a 3D grid system in which the whole domain is defined in two grids ($g = 1$ and 2, with a factor of 2 resolution difference along the three axes), and with smaller regions of the domain being defined in higher resolution grids ($g = 4, 5, \ldots, g_{max}$) with successive factor of 2 increases in resolution.

We should note that the characteristics of this adaptive grid are similar to the ones of the Cobra code (see, e.g., Falle and Giddings[5]; Falle and Raga[6]). However, it is not possible to evaluate the differences between the two codes as the details of the adaptive grid of the Cobra code have not been published.

3 Validation of the code

In order to validate the code, we compute a numerical model of a "starting jet" laboratory experiment. In the laboratory experiment, a 100 mJ Nd:YAG laser (with a 7 nanosecond pulse duration) is focused inside a 1 cm glass bubble. This glass bubble has an exit nozzle, which is directed away from the laser.

An approximately conical region surrounding the beam convergence cone (Fig. 2, top panel) is partially ionized by the laser pulse (with an ionization fraction of ~ 1 %) and heated to a temperature of $\approx 12,000$ K. This high pressure region then expands, pushing out a shock wave.

Part of this shock wave escapes through the exit nozzle, but a large region of the shock wave bounces aganist the wall of the glass bubble. This reflected shock wave rebounds again in the central region of the glass bubble, producing another outwards directed shock wave. Each time that the shock hits the surface of the bubble, a compression wave is ejected through the nozzle. For times $t \geq 100\,\mu s$, the gas that was heated by the laser pulse starts to leave the glass bubble through the nozzle, producing a very well collimated jet travelling into the surrounding atmosphere. The numerical simulation of this flow is shown in Fig. 2.

Figure 3 shows a comparison between the numerical simulation and shadowgraphs obtained from the corresponding laboratory jet experiment. It is clear that both the initial shock waves, as well as the high temperature (and low density) jet are very well reproduced by the numerical simulation. Interestingly, at times $t = 120$–$250\,\mu s$ the head of the jet develops a persistent vortex. We find that even the detailed structure of this vortex is well reproduced by the numerical simulation.

Some differences between the laboratory experiment and the numerical simulation are also evident in Fig. 3. For example, in the $t = 90\,\mu s$ frames, the numerical simulation shows a well defined jet beam and a vortex at the head

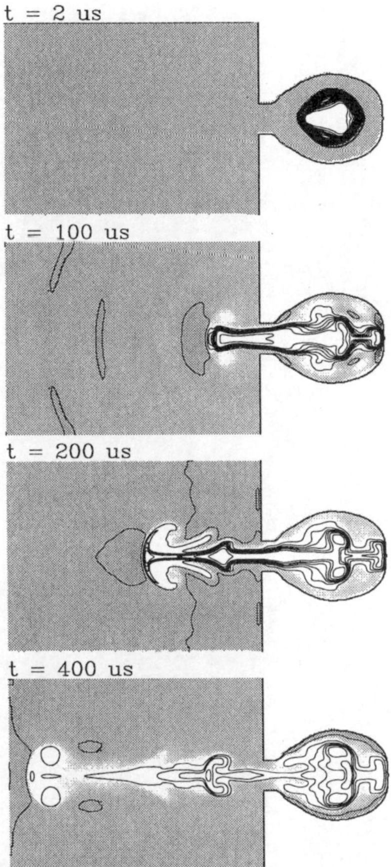

Figure 2. Numerical axisymmetric simulation of a jet produced by a laser-generated plasma bubble inside a glass bubble with an exit nozzle. The density stratifications for times $t = 2$, 100, 200 and 400 μs are shown with a linear greyscale and with contours with separations of 10^{-4} g cm^{-3}. The initially conical plasma bubble expands (top graph), sending out a shock wave that rebounds several times on the walls of the glass bubble. This results in the production of a series of compression waves that leave the glass bubble through the nozzle. At longer times, the air heated by the laser is ejected from the cavity, resulting in the production of a very well collimated jet (bottom graph). Each graph has a vertical spatial extent of 2 cm.

of the jet, while the jet beam is not visible in the corresponding shadowgraph. This absence of the jet beam in the shadowgraph is a result of the fact that

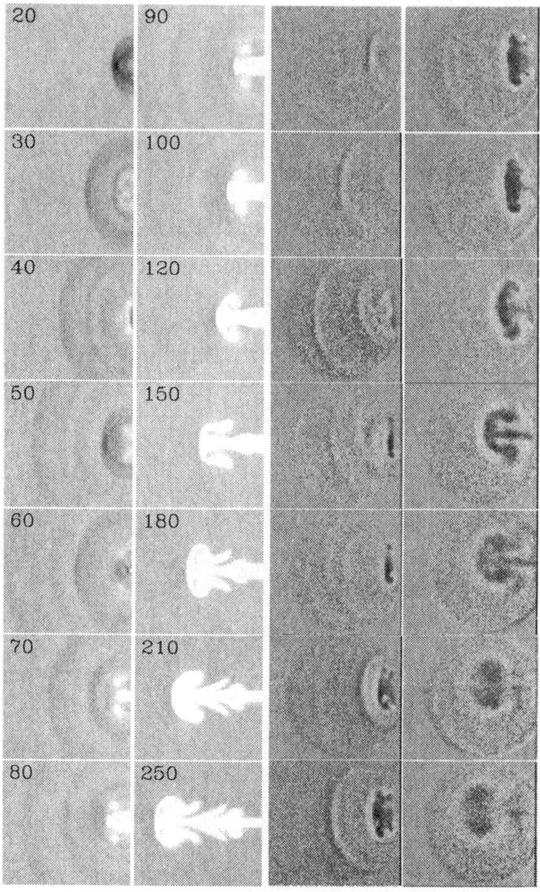

Figure 3. Density distribution predicted from the numerical simulation (first and second columns, also see Figure 2) and shadowgraphs obtained from the experimental jet (third and fourth columns) for times ranging from 20 to 250 μs from the laser pulse (the times in microseconds are written on the corresponding plots from the numerical simulation). Only the region outside of the exit nozzle is shown. Each of the graphs shows a region of 1.8×1.8 cm.

because of a shift in the optical components of the experimental apparatus, the region of the circular diaphragm close to the exit nozzle has fallen outside the CCD chip, and the jet beam is therefore outside of the image. The fact that the right hand side of the circular diaphragm falls outside of the imaged field is

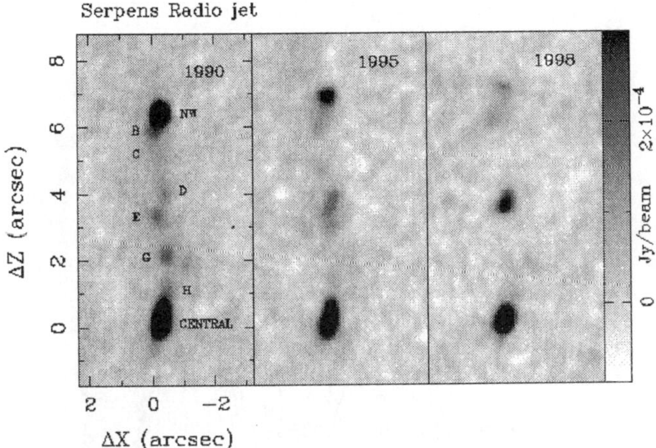

Figure 4. 3.6 cm radio continuum maps of the NW lobe of the Serpens radio jet obtained at three different epochs with the VLA (the Very Large Array interferometer in New Mexico, USA). The ordinate is aligned with a NW direction (PA= 48°.55). The proper motions and intensity variations of the knots along the jet can be clearly seen. The axes are labeled in arcesconds, with the zero point corresponding to the position of the outflow source. The identification of the radio knots was taken from Curiel et al.[9].

evident in many of the shadowgraphs shown in Fig. 3, and in order to compare them with the numerical simulation it is necessary to look at the positions of the features of the jet relative to the left edge of the circular diaphragm (which in all of the shadowgraphs falls within the CCD chip). Some of the differences observed between the numerical simulations and the experiment, however, are intrinsic differences, such as the lack of axisymmetry observed in the experiment at $t = 120\,\mu s$, which of course will never be reproduced by an axisymmetric simulation such as the one that we have computed.

4   Observations and models of a jet from a young star

Let us now discuss an application of our gasdynamic code to an astrophysical jet. We model the "triple radio source" in Serpens, which is a bipolar jet system ejected from a young star (Rodríguez et al.[7]; Snell and Bally[8]). This jet shows several knots, with high proper motions which are clearly seen by comparing radio maps obtained over a timespan of a few years (see Fig. 4, and Curiel et al.[9,10]).

We compute a numerical model of a jet with a sinusoidal variable ejection

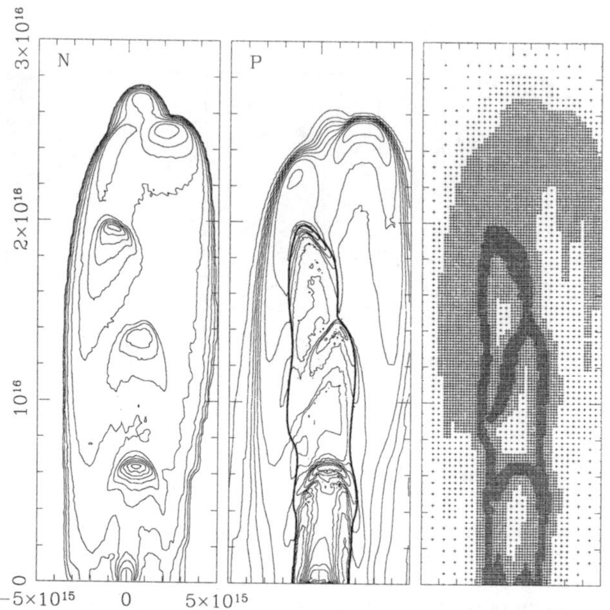

Figure 5. Column density (left, factor of $\sqrt{2}$ contours), pressure stratification on the central, $y = 0$ plane of the outflow (center, factor of 2 contours) and the points on this plane chosen by the adaptive grid algorithm (right), for the flow obtained after a 60 yr time integration. The thick contour in the pressure plot (center) delineates the contact discontinuity separating the jet from the environmental material. The axes are labeled in cm.

velocity of period $\tau_v = 10$ yr, half-amplitude $\Delta v = 80$ km s^{-1} and mean velocity $v_0 = 200$ km s^{-1}. The injection velocity has a direction at an angle of $5°$ from the z axis, initially lying on the xz plane, and then precessing around the z axis with a precession period $\tau_p = 50$ yr. We also impose an initial jet radius $r_j = 5 \times 10^{14}$ cm.

We choose an initial jet density of $n_j = 10^4$ cm^{-3} and a $n_{env} = 10^3$ cm^{-3}, uniform density for the surrounding environment. These parameters can be deduced from the radio observations of the Serpens triple radio source, using analytic models of jets from variable sources.

The calculation is done on a hierarchical, binary adaptive grid with a maximum resolution of 7.81×10^{13} cm (along the three axes). The maximum resolution is only allowed in the region of space occupied by the material originally coming from the jet, and the region filled by environmental material is resolved at most with a grid of 1.56×10^{14} cm spacing. The resulting grid

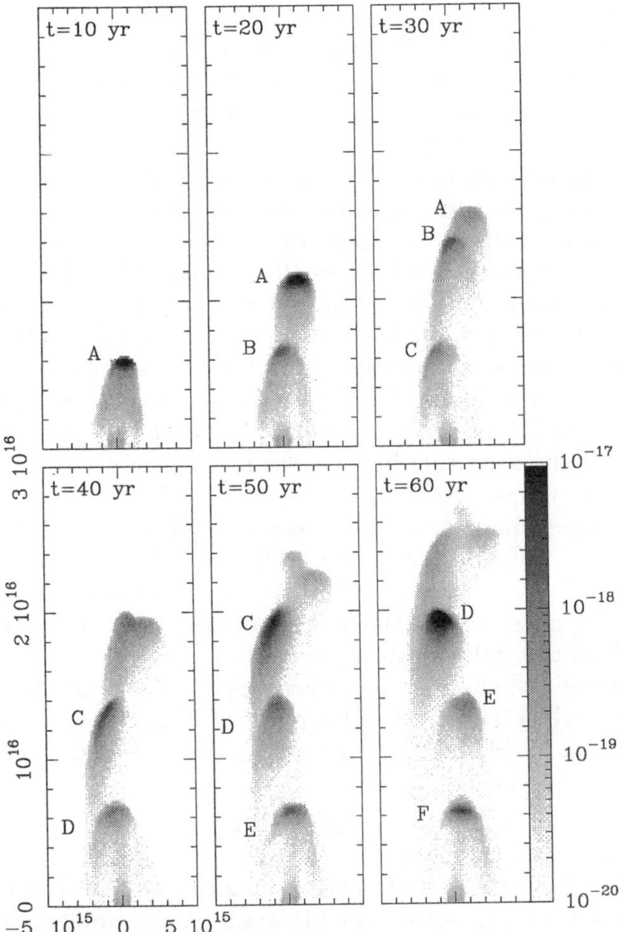

Figure 6. 3.6 cm radio continuum maps predicted from the numerical jet model for integration times from 10 to 60 years. The greyscales are logarithmic, with the range shown by the wedge (in erg s^{-1} cm^{-2} Hz^{-1} sterad^{-1}). The axes are labeled in cm. The successive working surfaces formed along the jet flow are labeled A through F, in order to identify the individual knots as they travel down the jet flow.

structure is illustrated in Fig. 5.

The flow stratification obtained after a $t = 60$ yr time integration is shown in Fig. 5, where we display the column density (integrated along the y axis) and the pressure stratification on the xz plane. The grid points on this plane (chosen by the adaptive grid algorithm) are also shown. In this figure, one can clearly see four separate working surfaces. The leading working surface has a complex structure which results from the merger of a number of knots that catch up with the leading bow shock of the jet.

Figure 6 shows the 3.6 cm, free–free continuum maps predicted from the numerical jet model for different integration times. The maps have been computed assuming that the xz plane coincides with the plane of the sky. In Fig. 6, we see the knots which are produced by the ejection velocity variability travelling down the jet flow and eventually catching up with the leading bow shock. The successive knots travel in different directions, as a result of the precession. It is clear that the knots show a complex time evolution.

Given the extreme complexity of the flow, it is hard to understand in detail the light curves obtained for the successive knots. As pointed out by Raga and Noriega-Crespo[11], the knots produced by a time-dependent ejection velocity have an intensity that first increases and then decreases as the knots travel away from the source.

In the present simulation, this effect is combined with the precession, which becomes more important at larger distances from the source[12]. As the knots diverge from each other at larger distances from the source[13], the trailing knots eventually cross the bow shock wings of previously ejected knots. These interactions lead to brightenings of the knots.

A more dramatic effect is seen when the trajectory of one of the knots intersects the leading bow shock of the outflow. This is seen for knot C in the $t = 50$ yr map and for knot D in the $t = 60$ yr map (Fig. 6).

5   Summary

We discuss the characteristics of the new, adaptive grid gasdynamic "yguazú-a" code. This code can solve 2 or 3D gasdynamic problems with explicit microphysics and coupled radiative transfer.

In order to test the code, we present a comparison between a numerical simulation and an experiment of a laser-generated starting jet. This comparison shows that the code reproduces the experimental results in a partially satisfactory way, indicating that the code does produce numerical solutions that correspond to correct integrations of the gasdynamic equations.

We then use this code to model an astrophysical jet. We have computed a simulation with parameters appropriate for the "triple source" jet in Serpens, which shows evidence for both a precession of the outflow axis and for a time-variability of the ejection velocity. We find that predictions of radio continuum maps from our numerical simulation agree qualitatively well with the observations. The predicted maps have a succession of bright knots which show large intensity variations over periods of a few years, also in agreement with the observed variability of the Serpens jet.

This jet simulation is the first astrophysical application of our recently developed and tested adaptive grid gasdynamic code.

Acknowledgments

This work was supported by the CONACyT grants 32753-E (AR), 27546-E (JC), 32531-T (RN-G), 27743-E and by the DGAPA (UNAM) grants IN 128098 (AR), IN 110998 (MV-M), IN 102796 (RN-G), IN 109297 (SC). LFR acknowledges support from CONACyT and DGAPA (UNAM). AR acknowledges support from a fellowship of the John Simon Guggenheim Memorial Foundation.

References

1. W. D. Harvey, W. W. Hunter, NASA Tech. Note D-7981 (1975).
2. B. Van Leer, ICASE report No. 82-30 (1982).
3. A. C. Raga, R. Navarro-González and M. Villagrán-Muniz, *RMxAA* **36**, 67 (2000).
4. A. C. Raga, G. Mellema, S. J. Arthur, L. Binette, P. Ferruit and W. Steffen, *RMxAA* **35**, 123 (1999).
5. S. A. E. G. Falle and J. R. Giddings, in *Numerical Methods for Fluid Dynamics* **4**, K. W. Morton and M. J. Baines, eds., 335 (Oxford, Clarendon, 1993).
6. S. A. E. G. Falle and A. C. Raga, *MNRAS* **261**, 573 (1993).
7. L. F. Rodríguez, J. M. Moran, P. T. P. Ho and E. W. Gottlieb, *ApJ* **235**, 845 (1980).
8. R. L. Snell and J. Bally, *ApJ* **303**, 683 (1986).
9. S. Curiel, L. F. Rodríguez, J. M. Moran and J. Cantó, *ApJ* **415**, 191 (1993).
10. S. Curiel, L. F. Rodríguez, J. F. Gómez, J. M. Torrelles, P. T. P. Ho and C. Eiroa, *ApJ* **456**, 677 (1996).
11. A. C. Raga and A. Noriega-Crespo, *AJ* **116**, 2943 (1998).
12. A. C. Raga, J. Cantó and S. Biro, *MNRAS* **260**, 163 (1993).
13. A. C. Raga and S. Biro, *MNRAS* **264**, 758 (1993).

A NONHYDROSTATIC METEOROLOGICAL MODEL (MÉSONH) APPLIED TO HIGH ANGULAR RESOLUTION IN ASTRONOMY: 3D CHARACTERIZATION AND FORECASTING OF OPTICAL TURBULENCE

ELENA MASCIADRI

Instituto de Astronomía, UNAM,
Apdo. Postal 70–264, 04510 México, D.F., México
E-mail: elena@astroscu.unam.mx

In this paper are presented the main results obtained by the adaptation of the non-hydrostatic meteorological model MésoNH, developed at CNRM (Centre National des Recherches Météorologiques) in Toulouse (France) to high angular resolution astronomical observations. Challenges and perspectives for future progress are discussed.

1 Introduction

At the present time, as a result of different international scientific collaborations, there are many projects for modern telescopes (diameter $\geq 8\ m$) and interferometers. Telescopes of this dimension could support ambitious scientific programs such as the study of active galactic nuclei (AGN), the detection and exploration of extrasolar planets and the study of the circumstellar environment of young stars.

However, the refraction index fluctuations of the atmosphere (optical turbulence: OT), introduce amplitude and phase perturbations in the wavefronts coming from the astronomical sources, thus reducing the image resolution to that of a telescope of only 10 cm in diameter. So, the potential angular resolution of some microarcseconds is, in the presence of atmospheric turbulence, reduced to a few arcseconds. In spite of this, ground observations are competitive with respect to the space-based ones because the financial investment is lower and the telescope's life is longer.

Actually, a considerable amount of progress has already been achieved in knowing and correcting the wavefront perturbations (adaptive optics techniques) and, in many cases, ground telescopes can obtain better resolution than space-based ones. At the same time, different techniques for the measurement of optical turbulence are known and they have been employed to characterize astronomical sites.

Little has been done with regard to numerical simulations related to meteorological models applied to the calculations of optical turbulence. Here we present the adaptation of the nonhydrostatic meteorological MésoNH model, developed at CNRM (Centre National des Recherches Météorologiques) in Toulouse (France), to the astronomical context. The optical turbulence was parameterized and included in the meteorological MésoNH model [1]. Many advanced modifications were provided and today the model can simulate, in a coherent way, many of the principal parameters characterizing the wavefront perturbations[2-5]. In this work I report the main challenges related to the numerical technique and, at the same time, the more recent results that have been obtained.

2 Challenges for a 3D characterization and forecasting of the optical turbulence

Today many different instruments dedicated to OT estimation exist. They are mainly based on integrated measurements (such as the optical instruments Scidar and DIMM) or *in-situ* measurements (such as balloon radiosoundings). Only numerical simulations can provide a 3D characterization and a forecasting of the OT in time. What are the critical points related to this technique?

2.1 Contribution of the vegetation in the optical turbulence simulation

In the feasibility study done on Cerro Paranal[2] we used a high horizontal resolution (400 m) orographic model coupled to the atmospheric MésoNH model to simulate the OT. Because of the lack of vegetation and a low roughness of the topography, this can be considered a good model of reality. The San Pedro Mártir Observatory (SPM), on the other hand, differs from Cerro Paranal in that there are tall trees, especially pine forest, all around the site with the 2.2 m and 1.5 m telescopes. It is evident that, from a dynamical point of view, the presence of tall trees affects the friction of the atmospheric flow over the ground and the level of turbulence produced will be greater than it would be over a bare mountain. Is this increase in turbulence activity quantitatively significant for our simulations? In order to give an answer we computed two simulations using different values of the roughness over the whole surface. In this way we parameterize the vegetation with a roughness value. In the first case a constant value of 1 m and in the second case a constant value of 10 cm were used. Although this description of the ground roughness is not realistic, the aim of this test was only to estimate the quantitative contribution of the vegetation parameter to the optical turbulence production. In a previous paper[3] we proved that the turbulence distribution is different in the two cases. Analyzing the temporal seeing evolution simulated above the SPM Observatory for the two configurations we computed a difference averaged seeing $\Delta\varepsilon$ over 1 hour of simulation of about 0.1 arcsec and it seems that the seeing is worse for larger values of the roughness. This proves that, for this particular case, it would be preferable to couple the orographic to a vegetation model and, from a general point of view, that the geographic site characteristics have to be well described in order to have a correct OT simulation.

2.2 Spatial model configuration

The dimension of the geographic surface analyzed is an important parameter for the numerical simulations. The larger the surface, the greater the necessary memory for the computation. For the first time[4] we made a test to estimate the impact of the size of the surface on the simulations. We studied the temporal evolution of the seeing simulated over the SPM Observatory. Two different surfaces were analyzed: 60 km $\times$ 60 km and 38 km $\times$ 38 km. The difference in seeing ($\Delta\varepsilon = 0.05$ arcsec) was very small. We concluded that, in this case, it would be preferable to use the small configuration if one is interested only in the SPM peak. On the contrary, if one intends to characterize the whole region of the park of San Pedro Mártir, it would be preferable to use the second configuration.

2.3 Optical turbulence integration along lines of sight other than zenithal

At the present time all the parameters (such as the seeing ε, the isoplanatic angle θ_{AO}, etc.) coded in the model are integrated along the zenithal direction. This is very useful because measurements obtained from optical instruments at an angle different from the zenith are systematically corrected in order to give the vertical C_N^2 (z) profiles[a]. In this way the simulations can be simply compared with the measurements. This correction consists in multiplying the integrated quantities by $\sec(\theta)$ where θ is the angle between the line of sight and the zenith. This operation is based on the hypothesis that the C_N^2 horizontal distribution is uniform. This is a strong hypothesis and we could show[3,4] that it is not necessarily true especially in the first 10 km where the orographic effects are more important. This is one reason why it would be interesting to introduce another configuration in the model so as to have maps of integrated parameters only for the position of the observatory but for several lines of sight. This new configuration could be interesting for a flexible-scheduling application, for the already existing telescopes. Moreover, the results obtained from these simulations could be fundamental for

---

[a]The C_N^2 parameter quantifies the OT in the atmosphere.

the calibration of instruments such as the Generalized Scidar that measures the C_N^2 along a line of sight and not in an atmospheric volume. The numerical problem related to the integration along lines of sight different from the zenith is not trivial, particularly because the atmospheric model levels are not horizontal. These levels are like sheets following the orographic model. Here we show the first results obtained with this modification of the code. To have a reference we carried out a 1 hour simulation using the classical configuration that we will call *site testing configuration* in the following. Fig. 1 shows the obtained seeing map. The thin black lines mark the altitude isolines of the orographic model, the color table the seeing values. One can observe that the seeing is higher over the mountain chain and lower over the plateau in the South–East part of the map. At the same time one can see that the turbulence is not uniform over this region. In the new configuration (which we will call *flexible-scheduling configuration*), we take the central point (SPM Observatory) as a fixed point and we integrate with respect to lines of sight different from the zenith. These lines have a sample of 10 degrees in the azimuthal angle ϕ and 5 degrees in the θ angle. The ϕ angle extends over 360 degrees and the θ angle over 40 degrees. The integration with lines of sight having $\theta > 40$ degrees is not significant because the theory of small perturbations is true only for small θ. There is no consensus in fixing the limit of validation of the theory. Roddier[9] showed that at about 60 degrees some saturation effects begin to appear, so one could consider this as a maximum limit. In any case we prefer to use this limit to ensure that the theory used is correct. Figure 2 (left) shows the seeing obtained for the SPM Observatory making the hypothesis that the optical turbulence is uniform over horizontal planes. One can observe a circular symmetry with respect to the center. The only factor that modifies the seeing value is the $\sec(\theta)$ factor, with ε monotonically increasing with θ. Figure 2 (right) shows the seeing obtained integrating the real optical turbulence (as shown in Fig. 1) over the whole region. One can observe that: (a) The seeing does not have circular symmetry, (b) the difference $|\Delta\varepsilon| = |\varepsilon_{i,FS} - \varepsilon_{i,ST}|$ of seeing between points of the same circle in the two different configurations increases with θ, (c) the difference $\Delta\varepsilon$ can be quantitatively quite large; for example, one can find differences of the order of 1.03 arcsec for $\theta = 40$ degrees; and (d) the difference $\Delta\varepsilon$ can be negative. Also along particular lines of sight, the seeing can be better than the one obtained for the previously symmetric case.

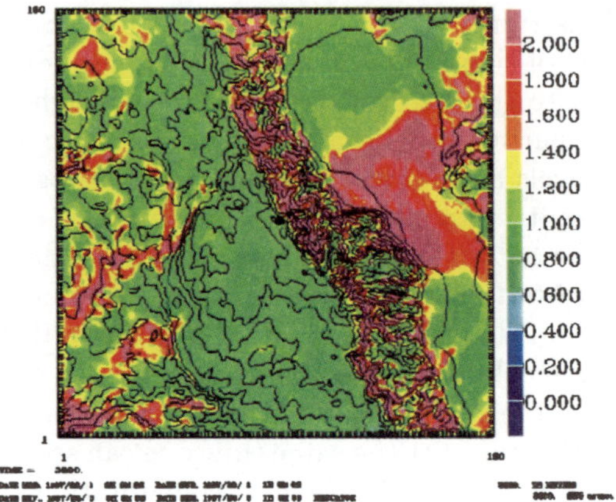

Figure 1. Map of the seeing obtained after 1 hour simulation above a region of 60 km x 60 km centered on SPM Observatory. The black lines mark the orography and the table of color represents the seeing intensity.

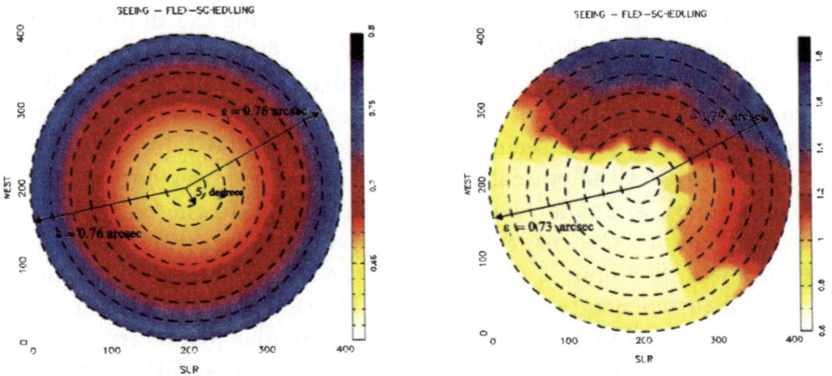

Figure 2. Polar map of seeing obtained after a 1 hour simulation having as fixed point the SPM Observatory. The center corresponds to the seeing obtained with an integration with respect to the zenith over the SPM. The circles drawn with dashed lines correspond to a seeing integrated at angles θ having a sample of 5 degrees. **Left:** uniform horizontal C_N^2 distribution. **Right:** real horizontal C_N^2 distribution.

Parameter [4 - 17.5 km]	Forecast (UT)	$\overline{\epsilon_R}(\%)$	$\epsilon_{R,\min}(\%)$	$\epsilon_{R,\max}(\%)$
Wind Intensity	6	8	4	25
"	12	14	5	36
Wind Direction	6	10.15	5.7	18.8
"	12	10.36	5.3	16
Abs. Temperature	6	0.2	0.1	0.28
"	12	0.3	0.2	0.37
Dew Point Temperature	6	1	0.14	1.9
"	12	1.18	0.17	23

Table 1. Relative error between the analysis and forecasts at 6 and 12 hours computed over a sample of one year. (31 N, 116 W) are the geographic grid point coordinates for which data were extracted from the ECMWF catalog.

2.4 Forecasting application

To validate the MésoNH model we initialize it with vertical meteorological parameter profiles (absolute temperature, dew point temperature, wind direction and intensity) obtained from the ECMWF (European Center for Medium Weather Forecasts) analysis or radiosoundings. These profiles are extrapolated over the whole analyzed orographic model. To forecast the OT one would initialize the MésoNH model with forecasted meteorological parameters (FMP). Thus, the confidence level of the OT forecasts depends on the quality of the FMP. In order to characterize statistically the latter we consider the analysis as the *real* meteorological conditions. We computed the relative error ($\overline{\epsilon_R}$) between the analysis and forecasts at 6 and 12 hours over the whole year 1997 for the San Pedro Martir site. In Table 1 the results obtained for each meteorological parameter are reported. One can observe that the temperature and the dew point temperature of the analysis and forecasts present an extremely good correlation ($\overline{\epsilon_R}(\%) \sim 1$ %). The wind direction and intensity, although not as good ($\overline{\epsilon_R}(\%) \leq 14$ %), present a good correlation too.

2.5 Seeing simulation in the millimetric range

The atmospheric effects on wavefront propagation depend strongly on the wavelength. The perturbations induced by the atmosphere over the wavefront are due to temperature and water vapor fluctuations and, consequently, to the refraction index fluctuations. The general expression for the refraction index is:

$$n - 1 = 77.6 \cdot 10^{-6} \frac{p}{T} - 5.6 \cdot 10^{-6} \frac{e}{T} + 0.375 \frac{e}{T^2} \tag{1}$$

where T is the absolute temperature, p the partial pressure of dry air and e the partial pressure of water vapor in millibar. The first two terms on the right-hand side arise from the polarization of gaseous constituents of the air. The third term arises from the permanent dipole moment of water vapor. In the visible range some strong approximations can be done and the expression of the refraction index can be reduced to:

$$n - 1 = 78.6 \cdot 10^{-6} \frac{p}{T} \tag{2}$$

The principal difference between the two expressions is the presence of water vapor which in the second case can be neglected. For astronomical observations at millimetric wavelengths the general form Eq. 1 more correctly describes the effects of the atmosphere on the wavefront. For millimetric observations the water vapor fluctuations are greater than the temperature fluctuations[8,10] in the sense that they produce the main contribution to the final perturbation. Therefore, the spatial–temporal fluctuations of the phase of a wavefront coming from a star are mainly due to water vapor fluctuations. At present, only the optical turbulence for observations in the visible range (C_θ^2) is parameterized in the MésoNH model. In order to use the model to simulate the turbulence that affects the observations in the millimetric range we made an analytic development of the refraction index variance and we coded it into the model. Here we do not describe the details of the analytic development. We recall only that:

$$C_N^2 = \left(\frac{\partial N}{\partial \theta} \right)^2 C_\theta^2 + \left(\frac{\partial N}{\partial q} \right)^2 C_q^2 + 2 \left(\frac{\partial N}{\partial \theta} \right) \left(\frac{\partial N}{\partial q} \right) C_{\theta q}^2 \tag{3}$$

where θ is the potential temperature, q is the specific humidity, C_θ^2 and C_q^2 are the temperature and humidity structure parameters and $C_\theta q^2$ is the covariance temperature–humidity structure parameter[7]. N is the refractivity and is equal to $(n - 1) \cdot 10^6$.

We did not yet compare simulations with measurements, but comparisons are planned. The MésoNH model would be, thus, a versatile tool for the site testing application in the visible and millimetric range.

3   Conclusions

In this paper we have exhibited some important issues related to simulations of the optical turbulence over an astronomical site. We showed the importance of using a vegetation model when tall trees are present around the site. We gave indications of how to optimize the analyzed surface dimension. We showed that the numerical technique can provide fundamental information related the 3D spatial distribution of the optical turbulence.

About the simulation of seeing in the millimetric range we can say that the ability to simulate the C_q^2 profiles depends on the ability to simulate the q profiles or, more specifically the gradient $\left(\frac{\partial q}{\partial z}\right)$. The specific humidity q is a parameter that has characteristic spatial–temporal fluctuations which are smaller than the fluctuations of the temperature θ. Moreover, the characteristic scale of the water vapor is only[10] 1 or 2 km. We think that a higher vertical resolution would be necessary to correctly simulate C_q^2 but at the same time, it would be probably sufficient to simulate only a part of the troposphere. Some questions remain unanswered: (a) Which data is it convenient to use to initialize the water vapor in the simulations? (b) Which technique is useful for measuring the water vapor? With respect to the latter problem, it appears that good results can be obtained with radiometer measurements[11] of the brightness temperature of the atmosphere.

4   Computer resources

The simulations presented in this paper were obtained using the Fujitsu VPP5000 supercomputer at Méteo France in Toulouse. It has 31 processors, each running at 9.6 Gflop/s. The MésoNH model can run currently in a parallelized version on the Fujitsu VPP5000. MPI interface programs are used for communications between the processors. The simulations presented in this paper extend over a surface of 60×60 km and an horizontal resolution of 400 m corresponding to $150\times150\times40$ grid points. The MésoNH model was recently installed on the Origin 2000 supercomputer at DGSCA–UNAM (Mexico). At present we are testing the performance of the latter supercomputer in order to verify the real possibilities of running such an atmospheric model (MésoNH) on an Origin 2000. This point is of particular interest because the Origin 2000 is a more widely available supercomputer that can be found in many institutions and universities.

Acknowledgments

A special thanks to Irene Cruz (TIM Project) and Fernando Angeles. This work was supported by CONACyT grant J32412E and DGAPA grant IN118199.

References

1. E. Masciadri, J. Vernin and P. Bougeault, *Astron. and Astrophys. Supp.* **137**, 185–202 (1999).
2. E. Masciadri, J. Vernin and P. Bougeault, *Astron. and Astrophys. Supp.* **137**, 203–216 (1999).
3. E. Masciadri and P. Bougeault, *Astronomical Telescopes and Instrumentation – SPIE Congress* , to be published (2000).
4. E. Masciadri, J. Vernin and P. Bougeault, *Astron. and Astrophys.*, submitted (2000).
5. P. Bougeault and E. Masciadri, *Rapport Technique Final: Simulation Numérique de la Campagne NOT 1995 – Convention ONERA 22.988/DA.B1/DC* , (1998).
6. R. Avila, J. Vernin and S. Cuevas Cardona, *PASP* **110**, 1106–1116 (1998).
7. J. C. Wyngaard, W. T. Pennell, D. H. Lenschow and M. A. LeMone, *J. Atmos. Sci.* **35**, 47–59 (1978).
8. M. L. Wesely, *Journal of App. Meteor.* **15**, 43–49 (1976).
9. F. Roddier, *Progress in Optics* **XIX**, 283–377 (1981).
10. A. R. Thompson, J. M. Moran and G. W. Swenson, *Interferometry and Synthesis in Radio Astronomy* , (1986).
11. C. L. Carilli, *Rad. Sci.* **34**, 817–840 (1999).

NUMERICAL SIMULATIONS OF INTERPLANETARY SHOCK WAVES USING ZEUS-3D

A. SANTILLÁN

*Cómputo Aplicado–DGSCA, Universidad Nacional Autónoma de México,
04510 México, D.F., México
E-mail: alfredo@astroscu.unam.mx*

J. A. GONZÁLEZ-ESPARZA AND M. A. YÁÑEZ

*Instituto de Geofísica, Universidad Nacional Autónoma de México,
04510 México, D.F., México
E-mail: americo@fis-esp.igeofcu,unam.mx*

We present preliminary results of a series of numerical simulations, in one and two dimensions with different resolutions (1024 and 3072 zones for the 1D case, and of 256×128 and 512×256 zones for the 2D case), of interplanetary shock waves using the magnetohydrodynamic (MHD) numerical code ZEUS-3D. Interplanetary shocks are produced by different perturbations associated with solar activity and propagate in the interplanetary medium throughout the solar wind. The objective of this study is to understand different physical characteristics of the origin and propagation of interplanetary shocks. The results of the numerical simulations will be compared with *in-situ* observations of interplanetary shocks by different spacecraft. The code ZEUS-3D has been tested and is used efficiently on the CRAY Y-MP and SGI Origin 2000 computers at UNAM's Supercomputer Center.

1   Introduction

The interplanetary space is immersed in the continual expansion of the solar atmosphere called solar wind. This expanding solar wind plasma and the interplanetary magnetic field carried by it, is the propagation medium for diverse types of waves and perturbations in interplanetary space. Solar flares, erupting filaments and coronal mass ejections (CMEs) are different manifestations of solar activity that can produce violent disturbances propagating in the corona and solar wind. Interplanetary shock waves (IPSWs) are the strongest perturbations in the solar wind and are related to many phenomena in interplanetary physics[1]. IPSWs play a very important role in solar–terrestrial relations and have been studied theoretically, observationally and numerically for over forty years. Based on previous experiences using the code ZEUS-3D in other astrophysical problems at UNAM's Supercomputer Center[2,3], we adapted the code to study IPSWs. We performed a parametric study of different perturbations in density, temperature and velocity injected into ambient solar wind to study their evolution. The aim of this investigation

Table 1. List of 1D numerical simulations. N_0, T_0, and V_0, denote the ambient solar wind density, temperature and velocity, respectively, at the inner boundary.

Case	ΔR (AU)	ΔN	ΔT	ΔV (km/s)	Δt (h)	W (ergs)	Resolution
1	2	$10N_0$	—	—	10	5.994×10^{30}	3072
2	2	—	$10T_0$	—	10	5.994×10^{30}	3072
3	2	—	—	$V_0 + 300$	10	3.960×10^{31}	3072
4	10	$100N_0$	—	—	30	1.029×10^{33}	1024
5	10	$10N_0$	—	—	30	1.618×10^{32}	1024
6	10	$2N_0$	—	—	30	8.472×10^{31}	1024
7	10	$1.5N_0$	—	—	30	7.990×10^{31}	1024
8	10	—	$100T_0$	—	30	1.029×10^{33}	1024
9	10	—	$10T_0$	—	30	1.618×10^{32}	1024
10	10	—	$2T_0$	—	30	8.472×10^{31}	1024
11	10	—	$1.5T_0$	—	30	7.990×10^{31}	1024
12	10	—	—	$V_0 + 1000$	30	3.185×10^{34}	1024
13	10	—	—	$V_0 + 500$	30	7.155×10^{33}	1024
14	10	—	—	$V_0 + 100$	30	2.131×10^{32}	1024
15	10	—	—	$V_0 + 50$	30	1.298×10^{32}	1024

is to illuminate the prime dynamic processes that govern shock evolution in the solar wind exploring different types of physical disturbances.

2   Numerical simulations using the ZEUS-3D code

The simulations are performed with the MHD code ZEUS-3D (version 4.2), which solves the three dimensional system of ideal MHD equations by finite differences on an Eulerian mesh[4]. The code can perform simulations in 3D but, in this work we restrict the discussion only to one and two dimensional simulations. The numerical experiments in this work illustrate the formation and propagation of shock waves in the solar wind based on the simplest possible quantitative model that contains the physics basic to these phenomena. In this model we neglected solar gravity and all magnetic effects, assuming the ambient solar wind to be a steady spherically symmetric, radial flow of an adiabatic fluid for which $\gamma = 5/3$. Under these assumptions, the solar wind physical characteristics vary with heliocentric distance (R): density $(\rho = \rho_0/R^2)$, pressure $(P_0 = P_0/R^{4.67})$, and sound speed $(C_S = C_{S_0}/R^{1.67})$. The simulations begin outside the critical point where the solar wind becomes supersonic $(R_0 = 0.18$ AU$)$ and thus do not address questions of how the wind

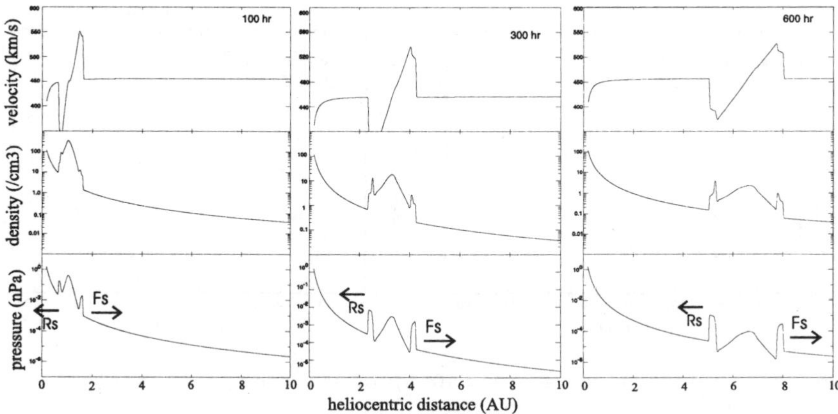

Figure 1. Temporal evolution of a density perturbation, case 4 ($\delta N = 100 N_0$). Plasma parameters: velocity, proton density number and thermodynamic pressure at three different times.

and perturbations themselves are initiated.

3 1D numerical simulations (HD)

Table 1 shows the list of the 1D numerical experiments we performed in the parametric study. We analyzed 15 cases; the second column shows the heliocentric range, in astronomic units (AU), that covers each simulation. The third, fourth and fifth columns show whether the injected perturbation into the ambient wind was in density, temperature or velocity, and the increment with respect to the ambient solar wind values. The next columns are the perturbation temporal duration, the total energy in the perturbation fluid ($W = \int 4\pi r^2 dr\{\frac{1}{2}\rho V^2 + \frac{3}{2}P\}$), and the mesh resolution for each simulation. For these experiments we used spherical geometry (R, θ, ϕ), injecting the flow from the inner boundary and leaving the outer boundary open. To produce the ambient solar wind we injected plasma at the inner boundary with initial values $V_0 = 410$ km/s, $N_0 = 128$ protons/cm^3, and $T_0 = 10^6$ K. Following Riley and Gosling[6] and Riley[7], the radial grid is filled with this wind and the algorithm is allowed to run in time until a dynamic equilibrium is achieved. After reaching equilibrium, we obtain at a heliocentric distance of 1 AU about the same solar wind averaged values measured *in-situ* by spacecraft ($V \sim 450$ km/s, $N \sim 8$ protons/cm^3). Now we discuss three cases of Table 1 that show how different perturbations have different evolutions.

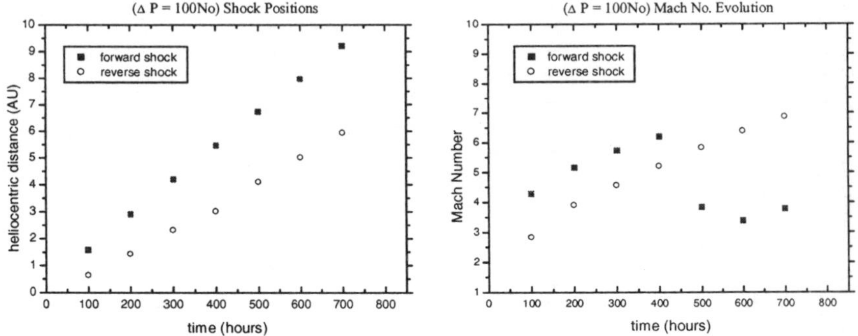

Figure 2. Shock evolution, case 4 ($\Delta N = 10 N_0$): (a) shock positions (forward and reverse) against time, and (b) shock Mach numbers against time.

3.1 Density perturbation (case 4: $\Delta N = 100 N_0$)

Figure 1 shows the plots of velocity, density and pressure at three different times (100, 300, 600 hours) after the injection of a perturbation with the same wind velocity and temperature, but much higher density ($\Delta N = 100 N_0$) during 30 hours. The perturbation produced a spherical high pressure region bounded by two shocks propagating in opposite directions with respect to each other: a forward (F_S) and a reverse (R_S) shock. Since the ambient wind characteristics vary with heliocentric distance there is a different evolution of both shocks with time. In order to study the local shock parameters we applied Rankine–Hugoniot relations around the shock vicinity and using averaged plasma values of the upstream and downstream regions we inferred the shock Mach numbers. Figure 2(a) shows the shock positions against time, where the two shocks are separating from each other following different evolutions. Figure 2(b) shows the shock Mach numbers against time, initially the forward shock was stronger but it began to lose strength, and around 400 hours after the injection the forward shock was decelerating. The reverse shock initially was weaker but it increased its strength continuously during the propagation, and after 500 hours it became stronger than the forward shock.

3.2 Temperature perturbation (case 8: $\Delta T = 100 T_0$)

Figure 3 shows plots of plasma parameters at three different times after the injection of a perturbation with the same wind velocity and density, but much higher temperature ($\Delta T = 100 T_0$) during 30 hours. Note that the total energy ($W = 1.029 \times 10^{33}$ ergs) of this perturbation is exactly the same than in the

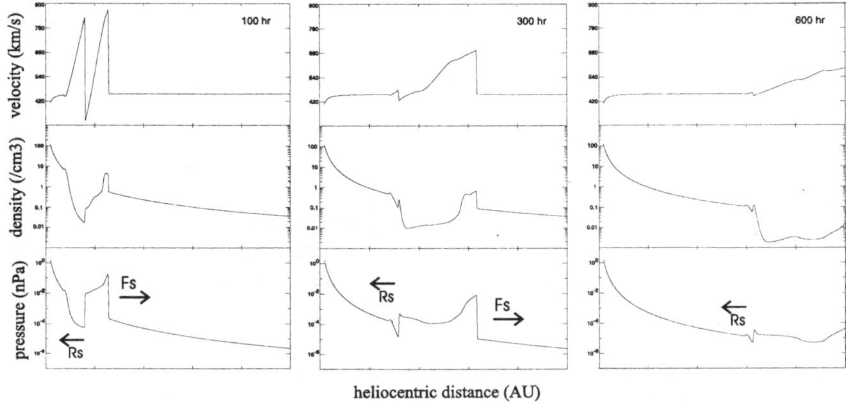

Figure 3. Temperature perturbation, case 8 ($\Delta T = 100T_0$). Plasma parameters: velocity, proton density number, and thermic pressure at three different times.

previous example (case 4), but the pressure increment is due to temperature instead of density. As in Fig. 1 the perturbation produces a spherical high pressure region bounded by a shock pair. However, in this case both shocks were stronger and faster, and therefore the reverse shock at the trailing edge of the compression region propagated into the inner boundary after the injection. The reverse shock appering in Fig. 3 is not the shock bounding the compression region, but it is a second reverse shock caused by the implosive wave in the rarefaction region. This perturbation in temperature follows a very different evolution than the perturbation in density of the previous case. Figure 4(a) shows the shock positions against time, note that the forward shock is much faster than in the previous example. Figure 4(b) shows the Mach number evolution against time. Although in both cases the perturbation internal pressure was the same, in this case ($\Delta T = 100T_0$) the shock strengths are about six times higher that in the previous density perturbation ($\Delta N = 100N_0$). The forward shock remains very strong, with a Mach number around 25 and overtakes the outer boundary after 500 hours. On the other hand, the second reverse shock diminishes in strength as the perturbation evolves with heliocentric distance.

3.3 Velocity perturbation (case 13: $\Delta V = V_0 + 500\ km/s$)

Figure 5 shows the plots of plasma parameters at three different times after the injection of a perturbation with the same temperature and density of the

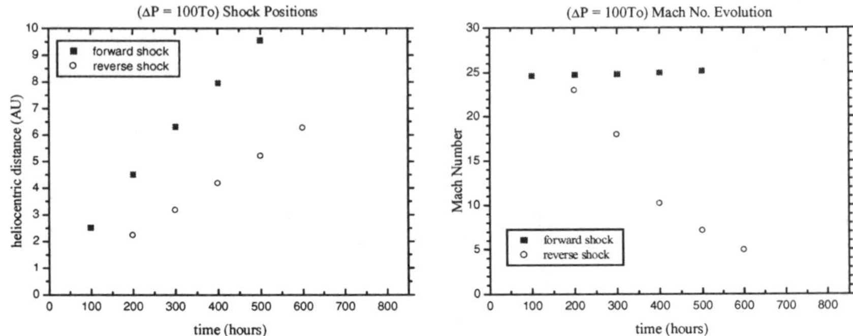

Figure 4. Shock evolution case 8 ($\Delta T = 10T_0$). (a) shock positions (forward and reverse) against time and (b) shock Mach numbers against time.

ambient solar wind, but much higher velocity ($\Delta V = V_0 + 500$ km/s) during 30 hours. The perturbation energy is about seven times higher than in the previous two cases ($W = 7.155 \times 10^{33}$ ergs), and the evolution was different. In this case, 100 hours after the jet injection there is a strong forward shock followed by a rarefaction region. 300 hours after the injection a weak reverse shock appears in the rarefaction region (similar to the Rs associated with the implosive wave in Fig. 3). 600 hours after the injection the forward shock is reaching the outer boundary and the reverse shock is increasing its strength. The analysis of the shock Mach numbers (not shown in this case) reveals that the forward shock was increasing in strength as propagated outward from the Sun with a Mach number around 15 and 17. The shock strenghts were very different in the three cases.

4 2D numerical simulations (HD)

Figure 6 shows an example of a 2D simulation of an interplanetary shock. The computational mesh covers a heliocentric range from 0.18 to 10 AU and longitudinal angular range from $0°$ to $180°$ with a resolution of 256×128 zones. Now the ambient solar wind was produced with initial values $V_0 = 740$ km/s, $N_0 = 80$ protons/cm^3, and $T_0 = 10^6$ K. The perturbation was injected into the ambient wind during 30 hours having the same temperature, but it was faster ($\Delta V = V_0 + 600$ km/s) and denser ($\Delta N = 4N_0$) than the ambient wind. The two snapshots (Fig. 6) show the perturbation evolution at 50 and 400 hours after being injected; the perturbation creates a strong interplanetary shock that tends to move radially away from the Sun. In the left panel the

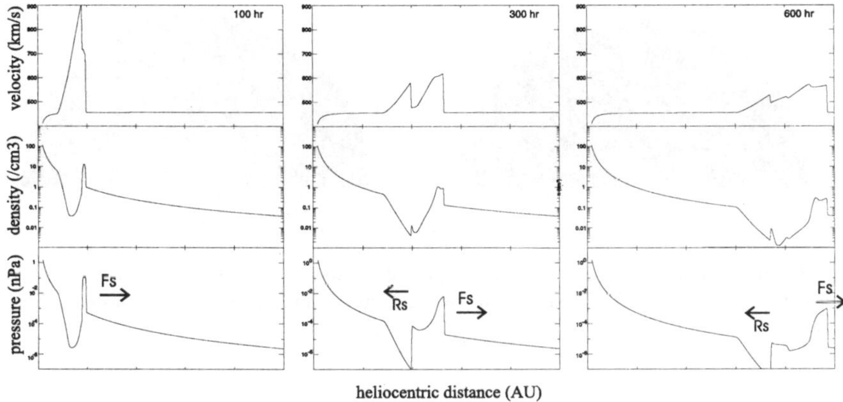

Figure 5. Velocity perturbation, case 13 ($\delta V = V_0 + 500$ km/s). Plasma parameters: velocity, proton density number and thermic pressure.

shock is located at 1 AU with a velocity of 800 km/s. Because of the great amount of energy that transports the perturbation, the shock will maintain the same conditions of velocity and density (800 km/s and 322 protons/cm^3, respectively) throughout the entire computational mesh. This is clearly seen in the right panel, where the shock is at 700 AU of the initial injection.

5 Conclusions and future work

We present the first results of a series of numerical simulations of IPSWs in one and two dimensions using the code ZEUS-3D. The simulations show that different perturbations (density, temperature, velocity) with similar fluid energy W produce different shocks with different evolutions. We should keep in mind that limited to one dimension (the radial direction), the simulation code predicts too strong an interaction between newly ejected solar material and the ambient wind because it neglects azimuthal and meridional motions of the plasma that help relieve pressure stresses. On the other hand the solar wind is a very structured magnetohydrodynamic fluid, and the conduction of heat by coronal and solar wind electrons leads to a flow that is not adiabatic. However, since the solar wind is a supermagnetosonic fluid and its dynamics is dominated by its dynamic pressure, these simple simulations help us to illuminate the prime dynamic processes of interplanetary disturbances. The next step in our study of interplanetary disturbances, is to analyze the heliocentric evolution, shock profiles and Mach numbers of the 1D experiments compar-

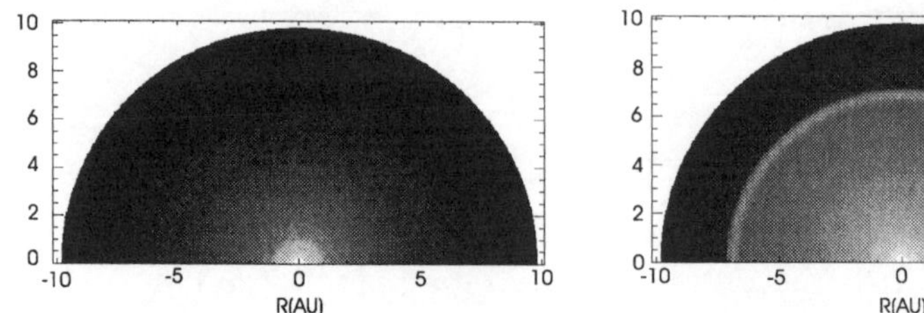

Figure 6. 2D numerical simulations of an interplanetary shock wave. The sequence shows the density (gray logarithmic scale) at two selected times (50 and 400 hours after the injection of a CME).

ing them with solar observations and *in-situ* solar wind measurements. We will use the 2D simulations to study the propagation of interplanetary shocks through structured ambient solar wind.

Acknowledgments

We are grateful to Dr. Riley and Dr. Wagner for useful discussions and motivation. The numerical simulations were performed on the SGI Origin 2000 at DGSCA–UNAM. This work was partially supported by DGAPA-IN115199 project.

References

1. R. G. Stone and B. T. Tsurutani, eds., *Collisionless Shocks in the Heliosphere: A Tutorial Review* (Geophysical Monograph 34, AGU, Washington D.C., 1985).
2. A. Santillán, J. Franco, M. A. Martos and J. Kim, *ApJ* **515**, 657 (1999).
3. A. Santillán, J. Kim, J. Franco, M. A. Martos, S. S. Hong and D. Ryu, *ApJ* **544**, in press (2000).
4. J.M. Stone and M. Norman, *ApJS* **80**, 753 (1992).
5. A. J. Hundhausen, in *Collisionless Shocks in the Heliosphere: A Tutorial Review*, R. G. Stone and B. T. Tsurutani, eds., (Geophysical Monograph 34, AGU, Washington D.C., 1985).
6. P. Riley and J. T. Gosling, *Geophys. Res. Lett.* **25**, 1529 (1998).
7. P. Riley, in *Solar Wind Nine*, S. R. Habbal, R. Esser, J. V. Hollweg and P. A. Isenberg, eds., (AIP Conf. Proc., 471, Woodbury, NY, 1999).

PHOTOIONIZING SHOCKS IN THE INTERSTELLAR MEDIUM

S. J. ARTHUR

Instituto de Astronomía, UNAM, Campus Morelia,
Apartado Postal 3-72 (Xangari), 58089 Morelia, Michoacán, México
E-mail: jane@astrosmo.unam.mx

A fast radiative shock in the interstellar medium (ISM) is a powerful source of ionizing photons. These photons are produced in the hot postshock cooling gas and can propagate both upstream and downstream. The photons travelling upstream encounter preshock gas and may produce an extensive precursor ionized region, while those travelling downstream influence the ionization and temperature structure of the recombination region of the shock. Supernova remnants (SNR) are a source of fast shocks in the ISM. In this work we investigate the influence a SNR has on the ionization of the ISM during its lifetime by means of hydrodynamic simulations of the evolution of remnants that include a detailed calculation of the nonequilibrium ionization state of the gas and radiative transfer of the radiation field produced in the postshock cooling region.

1 Introduction

Supernova remnants (SNR) are a particularly useful probe of the interstellar medium (ISM). These objects, visible in X-rays, optical emission and at radio wavelengths at various stages during their lifetimes, are composed of the ISM gas which has been swept up, compressed and heated by the blastwave from the supernova explosion. Assuming that the initial explosion is isotropic, any departure from sphericity and homogeneity in the morphology of the remnant and the distribution of emitted radiation tells us something about the nature of the ISM into which the remnant is expanding. SNR are relatively large objects and can probe the ISM over a wide range of lengthscales.

A supernova remnant traditionally has four major stages of evolution.[1] The first, free expansion stage, corresponds to early times in the remnant's life, when the supernova blast wave has not yet swept up more than the ejecta mass of interstellar material. Once a substantial amount of ISM material has been swept up, a reverse shock is driven back into the remnant, thermalizing the ejecta gas. The remnant enters the adiabatic phase and its expansion can be described by the self-similar Taylor–Sedov equations.[2] The hottest gas in the remnant is to be found in the centre, where the density is lowest. The coolest gas is found immediately behind the shock front. As the shock front decelerates, the temperature of the postshock gas becomes lower until it is in the range 10^5–$10^{6.5}$ K, when radiative losses start to become important. In the radiative phase, the dynamics of the shock front are affected by the cooling, and corresponding pressure changes, in the postshock gas. The final phase is when the pressure inside the remnant has dropped to almost the ambient pressure and the shock wave is carried forward by its own momentum.

This traditional evolution can be thwarted by inhomogeneities in the ambient interstellar medium. Type II supernovae are the result of the explosion of massive stars ($> 8M_\odot$). These stars have short lifetimes and are found in regions of star formation close to dense molecular clouds, where there are large density gradients. Moreover, massive stars modify their surroundings with their strong stellar winds and the H II regions created by their ionizing photons, leaving low density cavities in the ambient medium into which the SNR expand. In such an environment, an SNR will not have an adiabatic phase before the blast wave reaches the cavity wall and becomes radiative.

Radiative shocks in the interstellar medium are a strong source of ionizing photons. Cooling of hot gas behind a shock with velocity of a few hundred km s^{-1} produces X-rays and EUV photons. This radiation field travels both upstream and downstream and can affect both the ionization structure of the preshock gas, producing an ionized precursor (H II) region, and the ionization and cooling of the postshock gas.[3] For a constant velocity planar shock an equilibrium structure is produced, which can be divided into zones according to the ionization and cooling history of a parcel of gas as it travels through the shock wave. Not all of these zones are important in young supernova remnants, since the cooling times are much longer than the age of the remnant. Only the ionization zone (where preshock material suddenly finds itself in the hot postshock region and rapidly ionizes towards the collisional equilibrium state), the radiative zone (where the now-ionized gas begins to cool radiatively), and the nonequilibrium cooling zone (where cooling times are shorter than recombination times) need be considered. These zones produce a radiation field that consists of continuum (thermal bremsstrahlung, two-photon and free–bound) as well as line emission. This radiation can photoionize the region ahead of the shock wave. For the equilibrium shock model, the ionized region produced will be the corresponding equilibrium H II region.

Ionized precursor regions have been observed around a few SNR. Morse *et al.*[4] attributed diffuse emission, brightest in [O III], to a photoionized precursor ahead of the supernova shock in the Large Magellanic Cloud (LMC) remnant N132D. Bohigas *et al.*[5] observed decreasing [O III] and [O II] emission with increasing distance (up to 15 pc) from the shock front of the Cygnus Loop SNR, which they conjectured could be a photoionized precursor region. In both of these remnants, the estimated remnant age is too young for the SNR to be in the radiative phase if it were evolving in a uniform medium. Recently, Ghavamian *et al.*[6] have obtained evidence for a photoionized precursor around Tycho's SNR, where the photoionizing radiation is thought to be He II $\lambda 304$ photons produced in the immediate postshock region of a 2000 km s^{-1} shock.

However, in the case of SNR shocks we cannot expect either an equilibrium shock structure or an equilibrium precursor H II region. SNR shocks decelerate with time, hence the postshock region will be a complicated mixture of ionization regions and radiative zones from a continuous range of shock velocities. Furthermore, for young ($< 20,000$ yrs) SNR, the precursor region will not be an equilibrium H II region because the lifetime of the remnant is less than the recombination timescale of the precursor gas. A typical recombination timescale is $\sim 10^5/n_e$ yr, where n_e is the electron number density of the gas. In the region of N132D, the density is ~ 3 cm^{-3}, and the estimated remnant age is 3000 yrs, hence any precursor ionized region is likely to be far from ionization equilibrium. Similarly, in the case of the Cygnus Loop, typical ambient densities are ~ 0.1 cm^{-3} and the remnant age is $\sim$14,000 yrs.

In this paper, we examine the nonequilibrium H II regions formed around young supernova remnants. We perform hydrodynamic simulations, which include a treatment of the nonequilibrium ionization state of the gas, the calculation of the emergent ionizing spectrum from the postshock region and the radiative transfer of this ionizing radiation through the precursor region. In particular, we simulate the evolution of a supernova remnant within a low-density cavity—the current scenario for the N132D SNR[4]—to make a qualitative comparison with observations of that remnant.

2 Equations and method

We perform one-dimensional Lagrangian hydrodynamic simulations using a second order spherically symmetric Godunov scheme.[7] The advantage of a Lagrangian scheme is that there is no mixing of the ionization state between adjacent cells, hence the cooling, heating and ionization state can be followed accurately. The gas dynamic equations are

$$\frac{\partial}{\partial t}\begin{pmatrix} V \\ u \\ E \end{pmatrix} + \frac{\partial}{\partial m}r^2\begin{pmatrix} -u \\ p \\ up \end{pmatrix} = \begin{pmatrix} 0 \\ 2Vp/r \\ \mathcal{G} \end{pmatrix}, \tag{1}$$

where $V = 1/\rho$ is the specific volume, u is the velocity, $E = pV/(\gamma - 1) + \frac{1}{2}u^2$ is the specific total energy, p is the pressure, $\mathcal{G} = \mathcal{H} - \mathcal{L}$ is the energy source function (heating minus cooling rate) and γ is the ratio of specific heats. Here m is the mass coordinate, defined by $dm = \rho r^2 dr$.

In addition, we have the set of ionization equations,

$$\frac{dy_i}{dt} = U_{i-1}y_{i-1} - (D_i + U_i)y_i + D_{i+1}y_{i+1} \,, \tag{2}$$

where y_i is the fractional abundance of ionization state i (with associated charge $+z_i$) such that $\sum_i y_i = 1$ for each element. Here U_i is the total upward (ionization) rate from i and D_i is the total downward (recombination) rate from ion i. The total ionization rate consists of the collisional ionization rate, $n_e C_i^{\text{ion}}$, the photoionization rate, P_i^{ion}, and the charge exchange ionization rate due to collisions with ionized hydrogen and helium. The total recombination rate is comprised of the recombination rate, $n_e R_i^{\text{rec}}$, and the charge exchange recombination rate due to collisions with neutral hydrogen and helium. Here n_e is the electron density of the gas, where $n_e = n_I \sum_a \epsilon_a \sum_i y_i z_i$, where n_I is the number density of nuclei and ϵ_a the fractional abundance of element a. In this work we use abundances of hydrogen and the 12 next most abundant elements pertinent to the LMC,[8] that is, the metals are underabundant compared to abundances in the Solar neighborhood (this has important consequences for the cooling rate of the gas).

The ionization equations are coupled to the gas dynamic equations via the energy equation, since the net heating rate $\mathcal{G}$ is a function of temperature, T, radiation field, J, electron density, n_e, and ionization, y_i

$$\mathcal{G} = \mathcal{H}(J_\nu, y_i) - n_e \Lambda(T, y_i) \,. \tag{3}$$

Here, Λ is the cooling coefficient and the temperature is given by $T = p/(n_I + n_e)k$, where k is Boltzmann's constant.

The radiation field, J_ν, is composed of continuum and line emission produced by the radiating gas according to its temperature, density and ionization state. This radiation field interacts with the gas and is absorbed, emitted and scattered. This interaction is described by the equation of radiation transfer,

$$\frac{dI_\nu}{d\tau_\nu} = I_\nu - S_\nu \,, \tag{4}$$

where $I_\nu(\tau_\nu)$ is the specific intensity at frequency ν, and $S_\nu(\tau_\nu)$ is the radiation source function (defined as the ratio of emissivity to opacity). The optical depth τ_ν is the integrated opacity along the photon path. The mean intensity, $J_\nu(\tau_\nu)$, is the specific intensity averaged over all photon directions. This radiation field propagates both upstream, into the preshock gas where it is absorbed, thereby causing the gas to be heated and ionized, and downstream, where it affects the ionization state of the postshock gas.

We treat the ionization, the radiation, the energy and the gas dynamics as a fully coupled system. The timestep used for advancing the solution is the minimum of the hydrodynamic, cooling and heating timesteps over the whole computational grid. In practice, we also consider a large multiple (~ 500) of the ionization timescale, which is important in the region just behind the shock front. The set of ionization equations is solved implicitly, otherwise the calculation becomes prohibitively slow.

3   Observations of the LMC SNR N132D

Comparison of X-ray and optical images of the LMC SNR N132D show that the diffuse optical emission begins just outside the limb-brightened X-ray rim.[4] Analysis of this optical emission as a photoionized region gives an electron density of $n_e \sim 3$ cm^{-3} in this preshock gas.[4] From the X-ray data,[4,9] the X-ray flux can be fit by an electron density of ~ 15 cm^{-3}, an X-ray shell thickness of 2×10^{18} cm, and a temperature $T_e = 8.4 \times 10^6$ K. This corresponds to a shock of velocity ~ 800 km s^{-1} moving into an ambient medium of density ~ 3 cm^{-3}, consistent with the analysis of the diffuse optical emission. The radius of the X-ray remnant is ~ 11 pc and the thickness of the [O III] diffuse emitting region is ~ 1 pc. The X-ray rim can be interpreted as marking the position of the SNR shock, and the diffuse optical emission as being the photoionized precursor.[4] If a Taylor–Sedov model is adopted for the remnant, this leads to an inferred supernova energy that is too high (10^{52} ergs) if the remnant is assumed to have evolved in a constant density medium with $n_0 = 3$ cm^{-3}, given the implied shock velocity. The conclusion of Morse *et al.*[4] is that the remnant has evolved inside a low-density cavity, where $n_0 \sim 0.2$ cm^{-3}. Hughes,[10] on the other hand, postulated that the supernova explosion occurred in a cavity with density ~ 0.01 cm^{-3}.

4   Results

In this work we consider the expansion of a supernova remnant in a cavity of radius 10 pc and number density 0.3 cm^{-3}, where the surrounding ambient medium has a density of 3 cm^{-3}. These parameters were chosen with a view to modelling the LMC SNR N132D. The temperature in the ambient medium is set at 2000 K, which means that the gas is essentially neutral here. The cavity is in pressure balance with the surrounding medium. Upon reaching the cavity rim the shock velocity is $V_s \sim 1300$ km s^{-1} and the temperature in the swept-up cavity gas is high, $T_e \sim 10^8$ K. This gas emits hard X-rays, which do not produce much ionization in the surrounding denser ambient medium since the photoionization cross sections are small for these photon energies. When the shock hits the cavity wall a slower transmitted shock ($V_s \sim 650$ km s^{-1})

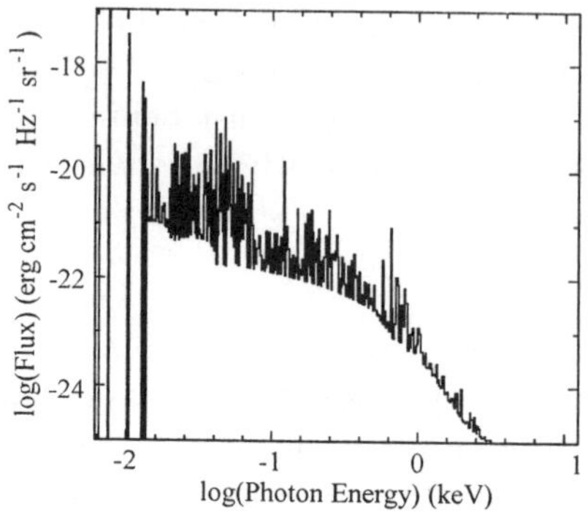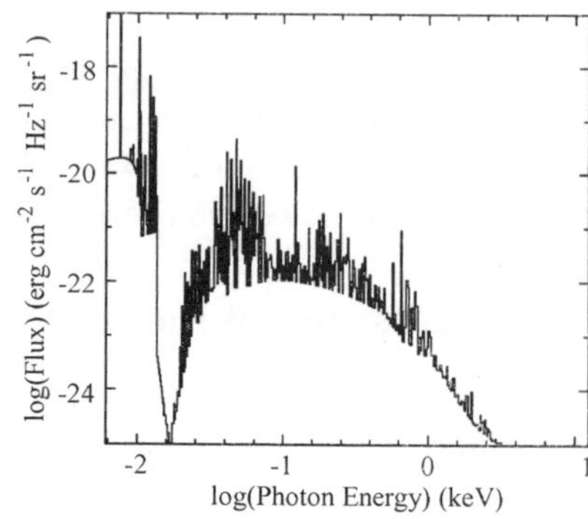

Figure 1. *Left:* Spectrum emerging from the shock into the precursor region. *Right:* Spectrum intercepted by gas a distance 0.8 pc from the SNR shock

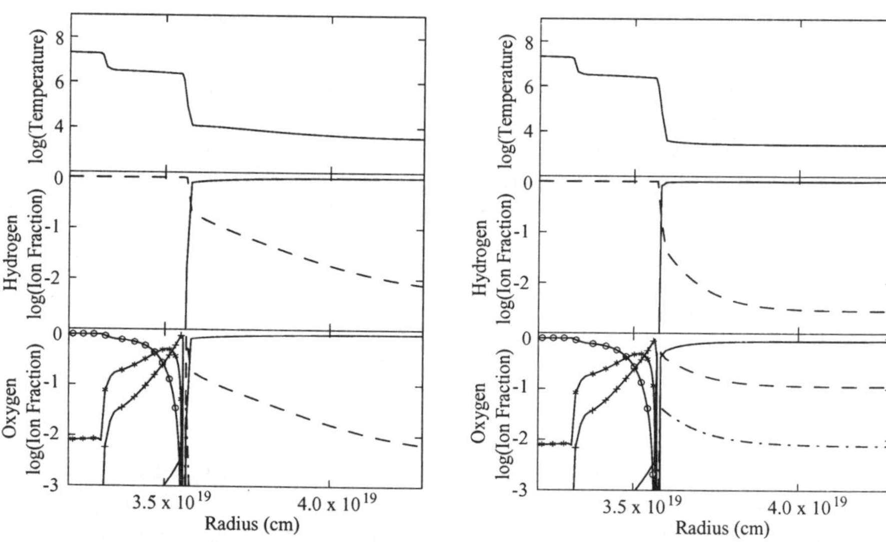

Figure 2. *Left:* Temperature (top panel) and hydrogen and oxygen ionization fractions (middle and bottom panels) in the immediate postshock and precursor regions. *Right:* Same as left figure but without charge exchange. *Key:* solid line—neutral species; dashed line—singly ionized species; dot-dashed line—doubly ionized species; line with crosses—O VII; line with stars—O VIII; line with open circles—O IX.

travels into the dense ambient medium. This shock decelerates as the remnant expands and the postshock gas starts to cool radiatively. By the time the region behind the transmitted shock has a thickness of ~ 1 pc, the shock velocity is below 400 km s^{-1}. The reflected shock that travels back into the cavity reheats the hot gas to temperatures $T_e > 10^8$ K.

In Fig. 1 (left) we plot a typical spectrum that emerges from the shock at a time when the region of swept-up dense material has a thickness of about 1 pc. At this point the transmitted shock has been travelling through the dense gas for some 3,500 yrs and has a velocity of ~ 370 km s^{-1}. We have made no attempt to match this spectrum to the observed *Einstein* flux in the 0.2–4.0 keV band and, in general, the flux in our spectrum is lower, by almost a factor of 10, than that observed. In this simulation, this part of the spectrum is produced by the extremely hot gas in the low-density part of the remnant. This is because the blast wave in the dense medium is not moving fast enough to produce this emission. In Fig. 1 (right) we show the spectrum intercepted by gas a distance 0.8 pc from the shock at the same time. We can see from this figure that the softer part of the spectrum is absorbed close to the shock and that only hard X-rays reach gas further out. The lowest-energy part of the spectrum is produced in the precursor region and is incapable of ionizing the gas.

The photoionizing radiation produces partial ionization in the precursor gas. In Fig. 2 we plot the ionization fractions of hydrogen and oxygen in the region immediately pre- and postshock. In the left hand figure charge exchange is included: the effect is to tie the oxygen ionization state to that of hydrogen. Hydrogen itself is only 10% ionized right next to the shock and this fraction falls off with distance. There is no O^{++} in the precursor region. The right-hand figure portrays the case when charge exchange is omitted. Although the postshock ionization structure is the same, in the precursor region oxygen achieves a few percent ionization to O^{++}. Charge exchange explains why oxygen is not more ionized in the precursor region.[11]

5 Conclusions

The X-ray emitting gas alone is not capable of producing the [O III] observed in the precursor region of the supernova remnant N132D. However, the simulations described in this paper assume that the remnant is expanding into essentially neutral gas at 2000 K. It is possible that if the remnant expands into an ionized region (such as the H II region formed by the progenitor star) then the ionization state in the precursor region will be higher. Oxygen will already be singly ionized in the H II region and it may be possible that the shock photoionizing spectrum can further ionize oxygen to produce the observed [O III] emission. We will perform new simulations where the remnant expands into the H II region formed by the progenitor star.

Morse *et al.*[4] also concluded that the X-ray emitting gas alone is not capable of producing the observed [O III] emission. They invoked the EUV spectrum of shocked dense cloudlets close to the X-ray rim of the remnant as the agent responsible for this emission. Alternative explanations could be the relic H II region itself (unlikely, since the progenitor is thought to be a B0 or later star), or the flash from the supernova. The N132D remnant contains many oxygen-rich filaments, which are seen in [O III] and [O II] emission. These filaments are not emitting X-rays,[4] however, radiative shocks moving through these O-rich filaments may be an important source of EUV photons.

In these simulations we were unable to reproduce the shock velocity ($\sim$ 790 km s^{-1}) required by the X-ray observations. This suggests that the cavity density used was too high and that the density postulated by Hughes[10] is more realistic. A lower cavity density would give a higher shock velocity when the remnant encounters the cavity wall. In the future, we will perform simulations of this scenario.

Acknowledgements

I would like to acknowledge financial support from DGAPA–UNAM (project number IN117799). I thank Will Henney for many useful discussions.

References

1. D. F. Cioffi, C. F. McKee and E. Bertschinger, *ApJ* **334**, 252 (1988).
2. L. I. Sedov, *Similarity and Dimensional Methods in Mechanics* (New York: Academic, 1959).
3. M. A. Dopita and R. S. Sutherland, *ApJS* **102**, 161 (1996).
4. J. A. Morse *et al.*, *AJ* **112**, 509 (1996).
5. J. Bohigas, J. L. Sauvageot and A. Decourchelle, *ApJ* **518**, 324 (1999).
6. P. Ghavamian, J. Raymond, P. Hartigan and W. P. Blair, *ApJ* **535**, 266 (2000).
7. S. K. Godunov, *Mat. Sb.* **47**, 271 (1959).
8. S. C. Russell and M. A. Dopita, *ApJS* **74**, 93 (1990).
9. U. Hwang, J. P. Hughes, C. R. Canizares and T. H. Markert, *ApJ* **414**, 219 (1993).
10. J. P. Hughes, *ApJ* **314**, 103 (1987).
11. J. P. Halpern and J. E. Grindlay, *ApJ* **242**, 1041 (1980).

NUMERICAL SIMULATION OF THE INTERACTION BETWEEN JETS AND SUPERNOVA REMNANTS

PABLO FABIÁN VELÁZQUEZ AND ALEJANDRO RAGA

Instituto de Astronomía, Universidad Nacional Autónoma de México,
Apdo. Postal 70-264, 04510 México D.F., México
E-mail: pablov@astroscu.unam.mx, raga@astroscu.unam.mx

We present results of numerical simulations performed to study the interaction between jets and Supernova Remnants (SNRs). Our aim is to explain the strange morphologies some SNRs exhibit in the radio-continuum, such as the case of SNR W50 in which a shell is interacting with the jets from the SS 433 source.

1 Introduction

At the end of their life, some stars can lose their thermal–gravitational equilibrium and die as a supernova explosion. A supernova is an event that occurs in extremely short timescales (the explosion itself takes just a few seconds, but the supernova can be visible for years after the explosion). A supernova remnant is composed of a shock wave, generated at the explosion time, the material ejected by the star, and the interstellar medium swept up by the supernova shock wave. In spite of the short lifetime of supernovae, their remnants can hold on for hundreds of thousand years, drastically modifying their environments and emitting energy in the whole electromagnetic spectrum, although the characteristic feature of SNRs is that they are strong sources of radio emission of nonthermal origin (synchrotron radiation).

Astrophysical jets are highly collimated flows which have been observed in many types of astrophysical objects (e.g., the nuclei of active galaxies, regions of stellar formation and compact objects).

Previously, there was no interest in studying the interaction between jets and SNRs, but recently there is observational evidence which shows that this kind of interaction is possible[1-3].

In this work we try to explain the strange morphology observed in the radio emission of the system composed by the source of relativistic jets SS 433 and the SNR W50. In order to do it we use a 2D adaptive grid code to simulate the encounter of a jet with a SNR shell.

SS 433 is a compact object (maybe a binary system) which emits two precessing jets in opposite directions and is located in the center of the supernova remnant W50. The precession cone has an initial half angle of $20°$ and the precession time is 164 days[4].

The SNR W50 shows a morphology at radiofrequencies which is far from being a spherical shell (which is the typical morphology of supernova remnants expanding into a homogeneous and isotropic medium) because it has two lateral extensions or lobes, in the East–West direction. Its angular sizes are $2° \times 1°$ which correspond to 104 pc $\times$ 52 pc assuming a distance of 3 kpc (according to the HI study of Dubner *et al.* [1]). These radio lobes are aligned with the precession cone axis of the SS 433 jets, such as is shown in X-ray observations[2]. Furthermore, there is optical emission which looks like filaments, located at the beginning of the radio lobes.

Several authors have given possible scenarios to describe the SS 433/W50 system, using analytical and numerical models, which can be divided into two groups. The models in the first group are based on the hypothesis that this system is a bubble which was formed by the interaction between the SS 433 jets and the surrounding interstellar medium or that the jets are expanding into an interstellar medium (ISM) swept up by the wind of a stellar partner of SS 433 (e.g., König[5] and Kochanek and Hawley[6]). On the other hand, the models in the second group consider that the elongated morphology of W50 is due to the encounter between the jets from SS 433 and the SNR shell[7,8].

In view of observational (radio, optical and X-ray) evidences which support the framework of a jet propagating inside of a SNR, we consider the models of the second group as the best possibility for describing this problem.

Our aim is to try to understand the process of interaction between jets and SNRs and to simulate the radio emission (synchrotron mechanism) in order to compare the models with observations.

2   The model

2.1   General considerations and initial conditions

As we mentioned in the Introduction, the direction of the jets of SS 433 precesses, close to the source, within a $20°$ half opening angle cone with a period of 164 days[4]. However, far from SS 433, the opening angle cone decreases to $10°$. This collimation of the beams could be a consequence of the interaction between the jet and SNR shocks, which generates secondary reflected shock waves along the symmetry axis.

The precession period is much shorter than the typical dynamic evolution time of the jet, which is defined as $t_{dyn} = R_l/v_j \simeq 140\ yrs$, where R_l and v_j are the mean radius of the eastern lobe and the jet velocity, respectively. Therefore, in our analysis we can model the SS 433 jet as two oppositely directed cones of $10°$ of half opening angle moving into the surrounding ISM.

The flow into the conical surfaces can be taken as continuous since at large distances from the center, the gaps between successive coils are expected to fill out by the velocity dispersion of the outflowing gas[8].

The calculation was carried out with the Coral code[9] on a 5-level binary adaptive grid with a maximum resolution of 2.44×10^{17} cm.

The jet was modeled as a filled cone moving with a velocity of 7×10^9 cm s^{-1} (which is close to the real speed of the SS 433 jets). As initial radius and length for the jet we take 1.25×10^{18} cm and as initial jet density and jet temperature we take 0.5 cm^{-3} and 5×10^7 K, respectively, after considering the X-ray work of Brinkmann et $al.$[10]

We take the analytical Sedov solution for a strong explosion, to initialize the density, velocity and pressure distributions of the SNR. After several tests, we have chosen an initial radius and explosion energy of 10 pc (3×10^{19} cm) and 4.5×10^{50} ergs (a typical energy for Type II supernova explosions), respectively.

Finally, for the interstellar medium we have chosen a density of 1 cm^{-3} and a temperature of 10^4 K.

2.2 Simulation of the synchrotron emission

Following the work of Clarke et $al.$[11] and Jun and Norman[12], the synchrotron emission intensity can be written as:

$$i(\nu) = C_1 \, \rho^{1-2\alpha} \, p^{2\alpha} (B \, sin\psi)^{\alpha+1} \, \nu^{-\alpha} \tag{1}$$

with C_1 being a constant, ρ the density, p the pressure, B the intensity of the magnetic field, α the spectral index parameter, ψ the angle between the direction of the magnetic field and the plane of the sky, and ν the frequency.

The intensity of the magnetic field can be modeled with the relation:

$$B \, = \, C_2 \, \rho^{2/3}, \tag{2}$$

where C_2 is a constant. Then, considering Eq. (2), fixing the frequency and assuming that the magnetic field is toroidal ($\mathbf{B} = B \, \hat{\theta}$), Eq. (1) can be rewritten as:

$$i \, = \, A_\nu \, \rho^{(5-4\alpha)/3} \, p^{2\alpha} (cos\theta)^{\alpha+1} \tag{3}$$

where θ is the polar angle which is related to ψ through $\psi \, = \, \pi/2 - \theta$ (if the angle ϕ between the jet axis and the plane of the sky is set to 0).

When the angle ϕ between the jet axis and the plane of the sky is nonzero, it is straightforward to show that Eq. (3) takes the form:

$$i \, = \, A_\nu \, \rho^{(5-4\alpha)/3} \, p^{2\alpha} (cos^2\theta + sin^2\phi)^{(\alpha+1)/2} \tag{4}$$

We check whether the gas has been shocked by computing the specific entropy for each computational cell and comparing it with the entropy of the unshocked material. We then set to zero the emission in the region filled with unshocked gas. For the W50 case, the spectral index α was fixed at 0.5, following the work of Dubner et al.[1] and we chose $\phi = 21°$ (according to Hjellming and Johnston[4]).

3   Results

We carried out several runs with different geometries to describe the SS 433 jet. Hollow conical jets and different aperture angles were tried. With this configuration, the obtained results are similar to Kochanek and Hawley[6], which do not describe well W50's morphology. So, we tried with a filled cone considering the initial conditions mentioned in §2.1.

From our simulation we can describe the interaction between a jet and a SNR as a process which can be divided in three phases or stages. In the first stage, the jet material propagates practically ballistically in the ISM which has already been swept up by the SNR shock wave. The jet generates a head with a double shock structure (the "working surface"), formed by a bow shock moving into the SNR gas, a reverse shock or Mach disk which decelerates the jet gas, and a contact discontinuity between them, separating the shocked SNR and jet gas. The working surface has a velocity v_{ws} (obtained assuming pressure equilibrium between the material entering through the bow shock and the Mach disk) given by:

$$v_{ws} = \frac{\beta(x) \; v_j \; + u_{SNR}(x)}{1 + \beta(x)} \tag{5}$$

where v_j is the jet velocity, u_{snr} is the velocity of the gas inside the SNR and $\beta = \sqrt{\rho_j(x)/\rho_{snr}}$ is the square root of the jet to SNR density ratio ($\rho_j(x) \propto x^{-2}$ where x is the distance along precession cone axis).

In the second stage, due to the fact that the working surface is moving faster than the SNR shock front, after a time t_e the jet catches up and collides with the SNR shell and begins to push and distort the shell (as is shown in Fig. 1, in the t=300 and 600 yr frames). t_e can be analytically determined considering that $v_{ws} = dx_{ws}/dt$ and integrating Eq. (5). We obtain that the jet encounters the SNR shock wave at $t_e = 530$ yrs, which is consistent with our simulation.

In the third stage, the jet propagates more quickly into the unperturbed ISM forming an elongated lobe. The lobe surface begins to exhibit a wavy structure (see Fig. 1, in the 1200, 1500 and 1800 yr frames). The origin of

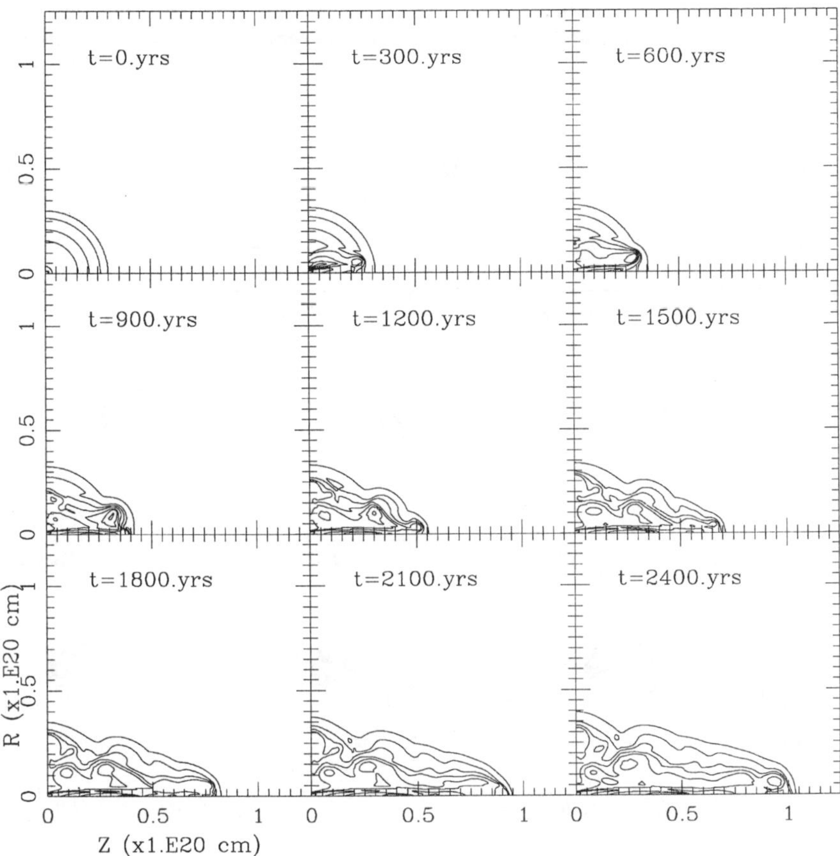

Figure 1. Density stratification of the interaction of a jet with a SNR shell. The successive contours correspond to factor of two in the density.

this wavy structure is beyond the scope of this paper and will be studied in future work[13]. We can say in advance that this structure is probably a result of the Kelvin–Helmholtz instability between the jet, cocoon (formed by jet, SNR and ISM shocked gas) and the external medium.

In the radial direction, the SNR maintains its spherical shape but in the last frames of Fig. 1 (t=1800, 2100, 2400 yrs), the cocoon fills up the SNR interior and seems to push out the SNR shell.

In order to compare these numerical results with radio observations, we simulated the synchrotron emission, following the steps described in §2.2.

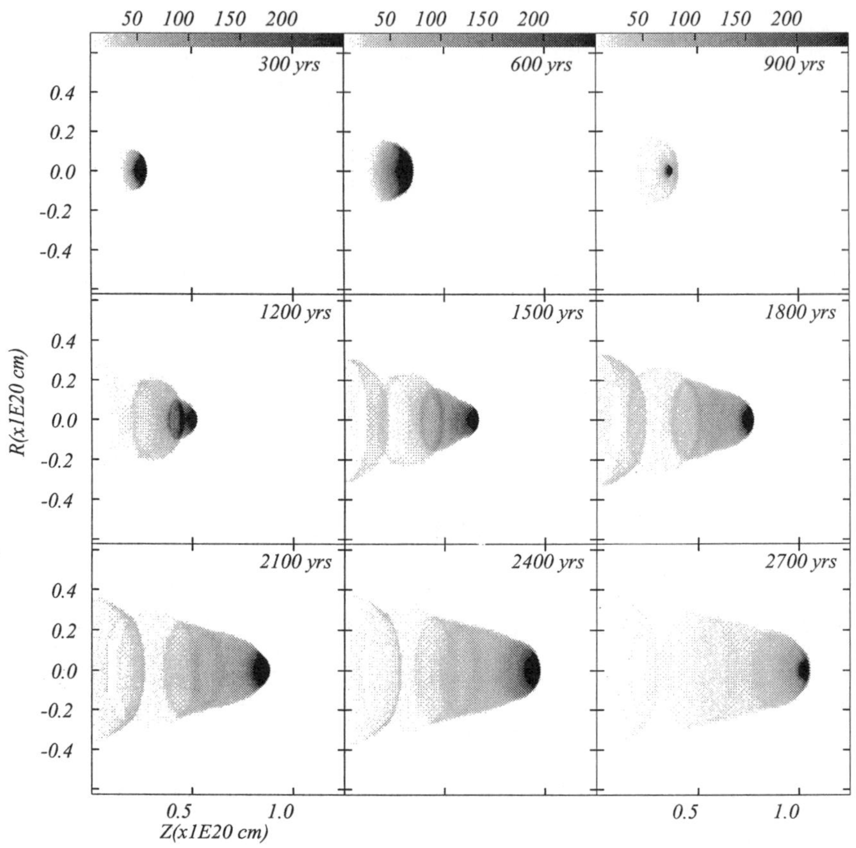

Figure 2. Normalized synchrotron emission of the interaction of a jet with SNR shell. The grays range is [0.05,0.250]. The angle of the jet axis with the plane of the sky is 21°.

Annular structures can be observed in our synchrotron maps (Fig. 2, t=1200, 1500, 1800, 2100, 2400 yr frames), which have separations of the order of the local lobe radius. A similar effect is observed in the real radio image (see the eastern lobe in Fig. 1(b) of Dubner et $al.$[1]).

4   Discussion

We show that a model in which a jet is interacting with a SNR shell success-fully explains the elongated morphologies of some SNRs, such as the case of the W50 SNR.

The simulated radio maps reveal the existence of annular structures, which are in good agreement with radio maps of W50.

A preliminary analysis carried out on our simulations shows that the jet dominates the dynamics of the lobe, after the encounter between the working surface of the jet and the SNR shock wave.

The cocoon seems to push and accelerate the SNR shell (in the radial direction). Several authors use the apparent unperturbed radius of the W50 shell, to determine the SNR age. If the cocoon is really pushing the shell of the remnant, the age will be overestimated. This effect and the origin of annular structures observed in the simulated radio maps, will be studied in detail in future work[13].

Acknowledgments

Pablo Velázquez is a postdoctoral Fellow of the Secretaría de Relaciones Exteriores de México, Plan Cuauhtémoc II. This work was supported by the CONACyT grants 32753–E and 27546–E.

References

1. G. Dubner, M. Holdaway, W. M. Goss and I. F. Mirabel, *Astron. J.* **116**, 1842 (1998).
2. S. Safi-Harb and H. Ögelman, *Astrophys. J.* **483**, 868 (1997).
3. B. M. Gaensler, A. J. Green and R. N. Manchester, *Monthly Not. Royal Astron. Soc.* **299**, 812 (1998).
4. R. M. Hjellming and K. J. Johnston, *Astrophys. J.* **328**, 600 (1988).
5. A. König, *Monthly Not. Royal Astron. Soc.* **205**, 47 (1993).
6. C. S. Kochanek and J. F. Hawley, *Astrophys. J.* **350**, 561 (1990).
7. W. J. Zealey, M. A. Dopita and D. F. Malin, *Monthly Not. Royal Astron. Soc.* **192**, 731 (1980).
8. K. Murata and N. Shibazaki, *Pub. Astron. Soc. Japan* **48**, 819 (1996).
9. A. C. Raga, G. Mellena and P. Lundqvist, *Astrophys. J. Suppl.* **109**, 517 (1997).
10. W. Brinkmann, H. H. Fink, S. Massaglia, G. Bodo and A. Ferrari, *Astron. and Astrophys.* **196**, 313 (1988).
11. D. A. Clarke, J. O. Burns and M. L. Norman, *Astrophys. J.* **342**, 700 (1989).
12. Byung-Il Jun and M. L. Norman, *Astrophys. J.* **472**, 245 (1996).
13. P. F. Velázquez, A. C. Raga, A. Costa, G. Dubner and D. O. Gómez, in preparation (2000).

THERMAL INSTABILITY IN A TURBULENT MEDIUM

A. GAZOL

Instituto de Astronomía, UNAM, Campus Morelia, Apdo. Postal 3-72, Xangari, 58089 Morelia, Mich., México

E. VÁZQUEZ-SEMADENI

Instituto de Astronomía, UNAM, Apdo. Postal 70-264, 04510 México, D.F., México

J. SCALO

Astronomy Department, University of Texas, Austin, TX 78712, USA

We investigate numerically the role of thermal instability (TI) as a generator of density structures in the interstellar medium (ISM), both by itself and in the context of a globally turbulent medium. Simulations of flows in the presence of the instability only show that the condensation process which forms a dense phase ("clouds") is highly dynamical. Final static situations, characterized by a bimodal or single peaked with a slope change density histogram (PDF), may be established, but the equilibrium is very fragile. Simulations containing the instability and various types of turbulent energy injection show that large-scale turbulent forcing is incapable of erasing the signature of TI in the density PDFs, but small-scale, stellar-like forcing causes the PDFs to transit from bimodal to a single-slope power law, erasing the signature of the instability. The third group of simulations are models of the ISM including the magnetic field, the Coriolis force, self-gravity and stellar energy injection. These simulations show no significant difference between the PDFs of stable and unstable cases, and reach stationary regimes, suggesting that the combination of the stellar forcing and the extra effective pressure provided by the magnetic field and the Coriolis force overwhelm TI as a density-structure generator in the ISM, TI becoming a second-order effect.

1 Introduction

The linear stability analysis for TI was first worked out in detail by Field[1], who clearly delineated the isobaric, isochoric and isentropic modes, and examined effects such as magnetic fields and conduction. This work has been subsequently generalized (e.g., Yoneyama[2]) and refined (see Wolfire *et al.*[3] for a recent discussion). In its simplest form, the isobaric mode of TI occurs when a compression leads to such enhanced cooling that pressure in the compressed region decreases. This leads to a picture of dense clouds in pressure equilibrium with their surroundings, the basis for the "two-phase" interstellar medium (ISM) model of Field *et al.*[4], which explains the existence of diffuse interstellar clouds. This conceptual framework was extended by Cox and

Smith[5] and McKee and Ostriker[6] to include a third, hot phase, but still assumes that clouds form by TI. We are interested in finding whether this sort of 2- or 3-phase model is relevant, even as a zeroth-order approximation, in a turbulent ISM, and, if not, what sorts of residual effects TI might have.

The idea that clouds form by TI in the ISM is often justified in terms of the near-constancy of the thermal pressure observed over a range of densities (0.1 to 100 cm^{-3}) (e.g., Myers[7]). However, closer inspection of Fig. 1 of Myers suggests that instead $P \sim \rho^{1/4}$ in that density range, and that, really, the effective exponent of this relation is a function of the density over the whole range of densities considered by Myers. Moreover, as pointed out by Ballesteros-Paredes *et al.*[8] (hereafter BVS99), *a constant thermal pressure does not imply dynamical equilibrium, as there are many other sources and sinks in the momentum equation for the ISM.* On the other hand, significant pressure imbalances between the hot and warm components of the local ISM have been reported by Bowyer *et al.*[9] A recent summary of (mostly observational) arguments against the specific three-phase McKee and Ostriker model is given by Elmegreen[10]. Moreover, the conclusion that pressure equilibrium is unlikely because of the frequency of either cloud collisions or shocks from supernovae or cluster superbubbles has been found independently several times in the literature (e.g., Stone[11], Heathcote and Brand[12], Bowyer *et al.*[9], Berghöfer *et al.*[13], Kornreich and Scalo[14]).

The basic question is whether dynamics of the gas (shocks, cloud collisions, or the turbulence itself) will disrupt incipient clouds faster than they can condense. Even within the context of linear stability analyses, there have been several suggestions that motions induced by the condensation process and other instabilities can supress TI or disrupt the structures formed by TI (e.g., Balbus and Soker[15], Murray *et al.*[16]). Thermal instability is not the only physical ingredient influencing the dynamics of the ISM; other equally important ones include self-gravity, the magnetic field, shear and the Coriolis force due to Galactic rotation and, of course, turbulence (operationally, the advection $[\mathbf{u} \cdot \nabla \mathbf{u}]$ term in the momentum equation). Elmegreen[17,18] (see also Passot *et al.*[19] for the two-dimensional case) has investigated the linear stability when all these processes (except turbulence, which is an intrinsically nonlinear phenomenon) are retained. The combined instability behaves very differently from TI alone, so the pure TI may represent an incomplete dynamical scenario even in the purely linear case.

What about thermal instability in a supersonic turbulent medium, like the gas in galaxies? (By "turbulent" we mean a disordered but not completely random velocity [and density] field covering a large range of scales.) In the ISM of the Milky Way there is certainly strong evidence for supersonic motions at all scales above at least 0.1 pc, in every sort of environment. Supersonic spectral linewidths are observed even in regions with no detectable internal star formation (e.g., the Maddalena molecular cloud complex; see Williams and Blitz[20]) and in diffuse mostly-H I clouds in which self-gravity is unimportant (known since the 1950s; see Heithausen[21] for a recent study of a subclass of such clouds), as well as in the more intensively studied star-forming molecular cloud structures.

Given all the above results, it is important to examine whether the traditional picture of a multiphase ISM structured by TI should be retained as a useful model. By "multiphase" we are referring to a medium containing various different thermodynamic *equilibrium* regimes, some of them possibly stable and others unstable, with the requirement that a multiphase medium involves fluid parcels undergoing a "phase transition" when transiting between these equilibria. The structuring of the density field by the multiphase nature of the medium should be reflected in multimodal density and temperature PDFs, with gas accumulating in the stable "phases".

2 The model

We use the numerical ISM model presented by Passot *et al.*[19], which uses a single-fluid approach to describe the ISM on a 1 kpc^2 plane on the Galactic disk, centered at the solar Galactocentric distance. In this model, various physical ingredients can be included at will, such as self-gravity, magnetic field, disk rotation, as well as model terms for the radiative cooling, a diffuse background radiation, and energy input due to star formation. The equations are solved in two dimensions at a resolution of 128^2 grid points (implying a spatial resolution of 7.8 pc) by means of a pseudospectral method. Further details concerning the model can be found in Passot et al.[19]

In our simulations, the thermal time scales are typically much shorter than the dynamical ones, implying that the turbulent motions are quasistatic compared to the background heating and cooling rates, and allowing for the establishment of thermal equilibrium (see Vázquez-Semadeni *et al.*[22,23]). The cooling function is taken as a piecewise power law of the temperature of the form $\Lambda = C_{i,i+1} T^{\beta_{i,i+1}}$ for $T_i \leq T < T_{i+1}$, where the values of the exponents and the coeficients are based on a reasonable fit to the radiative cooling curve of Dalgarno and McCray[24] for a fractional ionization of 10^{-4} at $T < 10^4$ K. A background heating is included to mimic that due to the diffuse UV radiation field or to heating by low-energy cosmic rays, and is taken as a constant $\Gamma_d = \Gamma_0$ everywhere throughout the duration of the simulations. The value of the constant is chosen in order to ensure that $T(\langle \rho \rangle)$ lies within the unstable range, with $\langle\rho\rangle = 1$. With the adopted power law behavior for these processes, the *thermal* equilibrium pressure has an effective polytropic behavior:

$$P_{eq} = \frac{\rho^{\gamma^e{}_{i,i+1}}}{\gamma}\left(\frac{\Gamma_0}{C_{i,i+1}}\right)^{1/\beta_{i,i+1}}, \tag{1}$$

where the (piecewise) *effective polytropic index* is given by $\gamma^e{}_{i,i+1} \equiv 1 - 1/\beta_{i,i+1}$. The isobaric mode of TI, characterized by a decrease in the pressure as the density increases, develops when $\gamma^e{}_{i,i+1} < 0$, corresponding in this case to $\beta_{i,i+1} < 1$ ($10 < T < 398$ K). Note that the condensation process induced by the instability does not stop when the density has crossed the values delimiting the unstable range, but at more disparate values of the density such that the pressures in the two phases are equal.

3 Results

In this section we sumarize our main results. For details concerning the simulations and a further discussion on the structure of the ISM see Vázquez-Semadeni *et al.*[25]

Simulations of the development of the TI alone, starting from small perturbations, show that the morphologies that arise depend quite sensitively on the proximity of the transition densities ρ_2 and ρ_3 (the values of the density bounding the unstable regime from above and below, respectively) to the mean density $\langle\rho\rangle$ of the medium. If the lower transition density ρ_3 is very close to the mean density, a large fraction of the mass remains in the warm (low-density) phase, and there is little mass available for forming the dense phase ("clouds"), which consists essentially of isolated, roundish structures. Conversely, simulations in which ρ_3 is significantly smaller than $\langle\rho\rangle$ evacuate more mass from the warm phase, and develop a transient filament network before the filaments are accreted into the actual peaks. These features can be seen in Fig. 1, where the PDF resulting from two simulations with different heating and cooling functions are plotted. The vertical lines denote the transition densities ρ_2 and ρ_3 for the two runs. In both cases, a peak in the PDF is observed just below the lower transition density, ρ_3. However, in the case of the run with $\Gamma_0 = 5$, ρ_3 is significantly different from the mean density, and this implies that a substantial amount of mass has to be transferred to the dense phase. As a consequence, a noticeable peak appears in the PDF largewards of the upper transition density, ρ_2. Instead, for the run with $\Gamma_0 = 3$, much of the mass remains in the low density phase, and there is no clear peak above ρ_2. Only a change in the slope of the PDF is seen there. Furthermore, the PDF for the run with $\Gamma_0 = 3$ seems to extend to much higher densities than that of the run with $\Gamma_0 = 5$. We interpret this as a consequence of the fact that the former reached a more advanced stage in the development of the instability than the latter, so that densities much higher than the transition value are reached as the densities in each phase approach the equilibrium values.

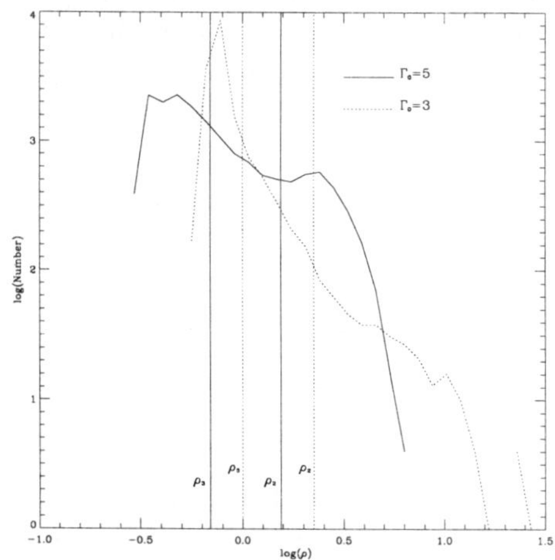

Figure 1. Density PDF for simulations with TI only

The condensation process which originates the "clouds" appears to be highly dynamical, with their boundaries being accretion shocks rather than static density discontinuities (see Vázquez-Semadeni *et al.*[25]). Although in principle the formation of the latter is possible, for realistic cooling functions, the equilibrium density in the cold phase (subscript "c") depends on that of the warm gas (subscript "w") as $\rho_c \sim \rho_w^{\gamma_w^e / \gamma_c^e}$, where γ_w^e and γ_c^e respectively denote the effective polytropic exponents of the warm and cold phases. For our fits, typically $\gamma_w^e / \gamma_c^e > 2$, so that small changes in the density of the warm gas require large changes in that of the dense gas, most likely implying the formation of shocks. Furthermore, if the density is continuous from one phase to the other, then the two phases must be mediated by larger pressure regions, which are expected to induce further motions. In summary, the static configuration is possible in principle, but appears highly unlikely.

The inclusion of energy injection processes ("forcing") has very different effects depending on the nature of the forcing. Random, large-scale forcing is not able to inhibit the development of the instability, and only distorts the

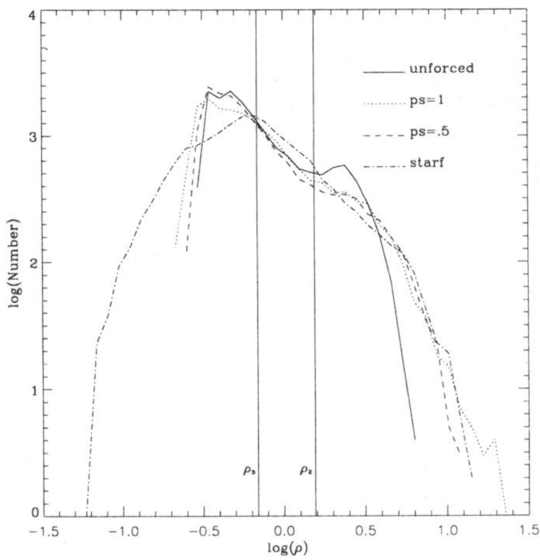

Figure 2. Density PDFs for forced simulations. Runs with $ps = 1$ and $ps = .5$ are large-scale forced runs and run 'starf' is a small-scale forced run.

structures it forms on time scales longer than the growth rates of the instability. In fact, large-scale turbulent modes naturally generate compressions, so the large-scale forcing actually *aids* the instability. Instead, small-scale, stellar-like forcing systematically injects energy within the structures formed by the instability, and is capable of reverting the compression of the dense gas and destroying the signature of the instability in the density distribution, pushing gas from the dense phase back into the unstable regime. In Fig. 2 we show the PDFs resulting from forced simulations, the PDFs corresponding to the two runs with large-scale forcing still show a significant slope change above the upper transition density. On the other hand, for the small-scale forced run the high-density peak spreads out, and in particular invades the unstable density region. However, the flow continues to be extremely compressible, as indicated by an ever-growing star formation rate and the inability to develop a stationary regime.

Finally, ISM-like simulations, including self-gravity, magnetic field, disk rotation and stellar like heating, with and without TI show little difference in their PDFs (Fig. 3). The main difference arises at low densities and it seems to be a consequence of the fact that the lower transition density for the unstable run is very close to the mean density and thus the warm phase is not strongly evacuated. Instead, for the marginal and stable runs, the lower transition densities are quite small (0.094 and 0.020). This small difference suggests that the combined effect of the stellar-like forcing, the magnetic pressure and the Coriolis force overwhelm the thermal pressure deficit in the unstable cases. Furthermore, all of the ISM cases reach stationary regimes, indicating that the presence of TI in the medium is of relatively minor importance in the overall cycle of such a regime, once the first generation of stars has formed. The likely disruption of clouds by Kelvin–Helmholtz instability[16] suggests that TI, even if it can occur in primordial galaxies, will be unable to form stable clouds.

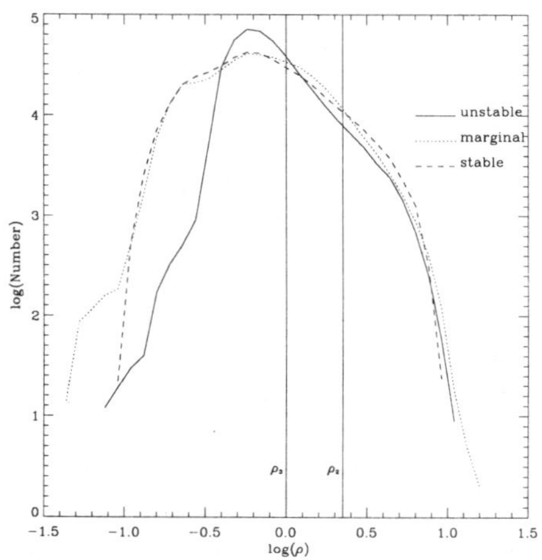

Figure 3. Density PDFs for ISM-like simulations.

We conclude that TI alone should not be expected to lead to static regimes (as also suggested by the recent simulations by Wada *et al.*[26], and that, in the presence of turbulence with small-scale driving, magnetic field and rotation, the resulting density distribution is substantially different from that due to TI alone. However, the largest scales (both over- and under-densities of sizes over 1 kpc) may indeed be formed preferentially by the combined thermal+gravitational+magnetic instability, as suggested analytically by Elmegreen[17,18] and numerically by Vázquez-Semadeni *et al.*[27], and confirmed in the nonmagnetic case by Wada *et al.*[26]

Acknowledgments

The simulations were performed on the Cray Y-MP 4/64 at DGSCA, UNAM. This work has received financial support from CONACYT grants 27752-E to EV-S and I32888-E to AG, as well as NASA grant NAG 5-3207 to JS.

References

1. G. B. Field, *ApJ* **142**, 531 (1965).
2. T. Yoneyama, *P. A. S. Japan* **25**, 349 (1973).
3. M. G. Wolfire, D. Hollenbach, C. F. McKee, A. G. G. M. Tielens and E. L. O. Bakes, *ApJ* **443**, 152 (1995).
4. G. B. Field, D. W. Goldsmith and H. J. Habing, *ApJ* **155**, L149 (1969).
5. D. Cox and B. W. Smith, *ApJ* **189**, L105 (1974).
6. C. F. McKee and J. P. Ostriker, *ApJ* **218**, 148 (1977).
7. P. C. Myers, *ApJ* **225**, 380 (1978).
8. J. Ballesteros-Paredes, E. Vázquez-Semadeni and J. Scalo, *ApJ* **515**, 286 (1999).
9. S. Bowyer, R. Lieu, S. D. Sidher, M. Lampton and J. Knude, *Nature* **375**, 212 (1995).
10. B. G. Elmegreen, *ApJ* **477**, 196 (1997).
11. M. E. Stone, *ApJ* **159**, 293 (1970).
12. S. R. Heathcote and P. W. J. L. Brand, *MNRAS* **203**, 67 (1983).
13. T. Berghöfer, S. Bowyer, R. Lieu and J. Knude, *ApJ* **500**, 838 (1998).
14. P. Kornreich and J. Scalo, *ApJ*, in press (2000).
15. S. A. Balbus and N. Soker, *ApJ* **341**, 611 (1989).
16. S. J. Murray, S. D. M. White, J. M. Blondin and D. N. C. Lin, *ApJ* **407**, 588 (1993).
17. B. G. Elmegreen, *ApJ* **378**, 139 (1991).
18. B. G. Elmegreen, *ApJ* **433**, 39 (1994).
19. T. Passot, E. Vázquez-Semadeni and A. Pouquet, *ApJ* **455**, 536 (1995).
20. J. P. Williams and L. Blitz, *ApJ* **494**, 657 (1998).
21. A. Heithausen, *A&A* **314**, 251 (1996).
22. E. Vázquez-Semadeni, T. Passot and A. Pouquet, *ApJ* **441**, 702 (1995).
23. E. Vázquez-Semadeni, T. Passot and A. Pouquet, *ApJ* **473**, 881 (1996).
24. A. Dalgarno and R. A. McCray, *ARAA* **10**, 375 (1972).
25. E. Vázquez-Semadeni, A. Gazol and J. Scalo, *ApJ* **540**, 271 (2000).
26. K. Wada, M. Spaans and S. Kim, *ApJ*, in press (2000).
27. E. Vázquez-Semadeni, T. Passot and A. Pouquet, *RMAA Ser. Conf.* **3**, 61 (1995).

TURBULENT DISSIPATION IN THE INTERSTELLAR MEDIUM IN THE PRESENCE OF DISCRETE ENERGY SOURCES

VLADIMIR AVILA-REESE AND ENRIQUE VÁZQUEZ-SEMADENI

Instituto de Astronomía, Universidad Nacional Autónoma de México,
Apdo. Postal 70-268, 04510 México, D.F., México
E-mail: avila,enro@astroscu.unam.mx

We explore the dissipative ability of turbulent compressible flows that model the interstellar medium (ISM) at intermediate-to-large scales. The main feature of our simulations is the (realistic) way in which the turbulent kinetic energy is injected to the fluid: around the star formation sites, the gas is accelerated radially away from the "stars", acquiring a well-defined final velocity difference u_f over a characteristic length scale l_f. We study the dependence of turbulent dissipation on these two quantities. The spatially scattered, small-scale nature of this forcing gives rise to the coexistence of both forced and decaying turbulent regimes within the same flow. In the forced case, the global dissipation time is proportional to $(l_\mathrm{f}/u_\mathrm{f})/u_\mathrm{rms}$, where u_rms is the rms velocity dispersion of the flow. The kinetic energy injection and dissipation rates are very close, implying that most of the turbulence is dissipated near the localized input sources. In the decaying regime, the kinetic energy decays as a power law in time, with an exponent ~ -0.8. Our results, if applicable to the vertical direction in the Galactic disk, are consistent with models of galaxy evolution in which large-scale star formation is self-regulated by an energy balance in the vertical component of the gaseous disk. On the other hand, our results do not support galaxy formation models in which the stellar energy injection in the disk is required to self-regulate star formation and reheat the gas at the level of the whole cosmological halo (of typical sizes 15–20 times larger than the optical galaxy).

1   Motivation

The thermal state and structure of the large-scale interstellar medium (ISM) in disk galaxies seem to be controlled by gasdynamical processes rather than thermal conduction as occurs in the thermally regulated three-phase model of e.g., McKee and Ostriker[1]. Indeed, turbulent, and probably magnetic and cosmic ray pressures, are at least three times larger than the thermal pressure (e.g., Boulares and Cox[2], McKee[3], Lockman and Gehman[4], Ferrara[5], Norman and Ferrara[6]). Therefore, in order to study the "metabolism" of the large-scale ISM in disk galaxies, thermohydrodynamic models and simulations should be used.

A crucial ingredient in these thermohydrodynamical models is the star formation (SF) cycle, where the main driver of this cycle seem to be the large-scale flows influenced by the disk gravitational potential. In turn, the large-scale SF cycle plays a key role on the processes of galaxy formation and evolution. A major question related to galaxy formation and evolution is whether the SF rate is self-regulated (by negative feedback) either at the level of only the disk ISM or at the level of a hypothetical intrahalo medium in hydrostatic equilibrium with the cosmological dark matter halo which extends to 15–20 times the optical radius of the galaxy. Since the *kinetic energy dissipation rate in the turbulent ISM* could be the main factor in determining the efectiveness of large-scale SF feedback, its study is required in order to shed light into this question.

Motivated by the above mentioned question, we have studied the dissipation of turbulence in compressible MHD fluids that resemble the intermediate-to-large-scale ISM in the solar neighborhood in our galaxy, paying special attention to the mechanism of driving the turbulence (Avila-Reese and Vázquez-Semadeni[7], hereafter AV00). Previous simulations on dissipation in compressible MHD fluids have focused on molecular-cloud-like (isothermal) regimes (Mac Low *et al.*[8], 1998; Stone *et al.*[9], Padoan and Nordlund[10], Mac Low[11]), and with the turbulence driven in Fourier space by large-scale random velocity perturbations whose amplitudes are selected as to maintain the global kinetic energy E_k constant in time. As a result, kinetic energy is injected everywhere in space (a "ubiquitous" injection). Instead, in the ISM the spheres of influence of the input sources are relatively small compared to the typical scales of the global ISM, of short durations, and located at discrete sites. In AV00, an injection mechanism based on self-consistent stellar winds was implemented. Here we briefly describe the method and some of the results and implications presented in AV00.

2   The method

The numerical model presented in Vázquez-Semadeni *et al.*[12,13] and Passot *et al.*[14] was used. The MHD equations, including the internal energy conservation equation, are solved in 2D in the presence of self-gravity, the Coriolis force, and model terms for radiative cooling, background heating, stellar winds and large-scale shear. The numerical technique is pseudospectral, requiring periodic boundary conditions. A relatively low resolution (128^2 grid points) was used in order to cover a reasonable range of parameters. Our main aim is to gain a general insight on the dissipative properties of the ISM, not to obtain quantitative results on the dynamics of the ISM; this last point will be treated in future works using other numerical schemes. The physical variables of the fluid were chosen in such a way the simulations resemble the ISM in the plane of the Galaxy near the solar neighborhood and at scales $\sim$10–1000 pc.

In our simulations, an "energy input source" is turned on at grid point $\vec{x}$ whenever $\rho(\vec{x}) > \rho_c$, and $\vec{\nabla} \cdot \vec{u}(\vec{x}) < 0$. Once SF is turned on at a given grid point, it stays on for a time interval $\Delta t_s = 6.5$ Myr, during which the gas receives an acceleration $\vec{a}$ directed radially away from this point. The input sources are extended spatially by convolving their spatial distribution with a Gaussian of width l_f. For the turbulent fluid, l_f is the forcing scale. At this scale the acceleration $\vec{a}$ produces a velocity difference $u_f \approx 2\vec{a}\Delta t_s$ around the "star" (the velocity at which turbulence is forced at the l_f scale). Both l_f and u_f are free parameters and we explore how dissipation depends on them running several simulations.

3 Dissipation in driven turbulence

The fiducial simulation in AV00 includes several interstellar "ingredients": self-gravity, heating and cooling, shear, magnetic fields, and, of course, the SF prescription. The values used for l_f and u_f are 30 pc and 30 km/s, respectively, a compromise between parameters representative of expanding H II regions and supernova (SN) remnants. In this type of run, the thermal pressure behaves very closely to a piecewise polytropic pressure ($P \propto \rho^\gamma$) (Vázquez-Semadeni et al.[13]) with an *effective polytropic exponent* γ which adopts different values between 0 and ~ 1 in different density ranges. In Fig. 1, images of the density field at times 14.3 Myr and 144.3 Myr are shown. At the earlier time, small shells are seen, whose sizes are comparable to the sphere of influence of the energy sources (of size l_f). At the later time, the shells are larger; their larger sizes are a consequence of the inertial motions induced by the forcing and of the induced-SF events in the shells.

Thanks to the "wind" SF prescription used, the kinetic energy injection rate $\dot{E}_k^{\text{in}}$ can be measured directly in the simulation. Using the measured values of E_k and $\dot{E}_k^{\text{in}}$, the kinetic energy dissipation rate $\dot{E}_k^{\text{diss}}$ can then be calculated from $dE_k/dt = \dot{E}_k^{\text{in}} - \dot{E}_k^{\text{diss}}$. Remarkably, $\dot{E}_k^{\text{in}}$ and $\dot{E}_k^{\text{diss}}$ are always very close to each other (note that the regime is not stationary). Since the energy injection occurs at very localized and scattered sites, while the dissipation rate is computed over the whole flow, their highly similar evolutions suggest that *most of the dissipation occurs at or near the injection regions*, and dominates the global dissipation rate. Only in this case can the global dissipation rate track the locally-originated injection so closely. In turn, this implies that only a small fraction of the energy injected at the sources "escapes" to more distant regions. We refer to this as the "residual" turbulence.

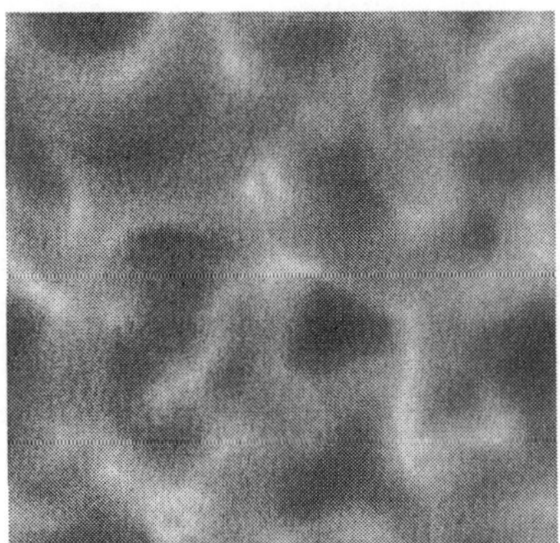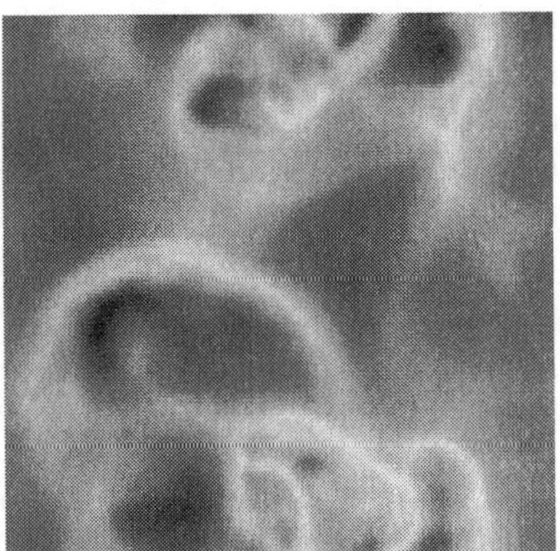

Figure 1. Images of the log of the density field of run ISM at $t = 14.3$ Myr (*left*) and at $t = 144.3$ Myr (*right*). The box size is 1 kpc and the resolution is 128 grid points per dimension. Star formation (SF) started at $t = 5.2$ Myr in this run. In the left panel small expanding shells are seen, still in their initial phases. The trapezoidal void near the middle was not formed by stellar activity, but by the turbulent initial conditions. At the later time, large shells of up to ~ 500 pc are seen. These are due to induced SF in the shells.

An important finding of AV00 is that in the case of small-scale, spatially-intermittent forcing, *forced and decaying regimes coexist in the same flow*, each one probably with its own characteristic dissipation time scale. The global dissipation time ($t_{\rm d} = E_{\rm k}/\dot{E}_k^{\rm diss}$) measured in the fiducial simulation fluctuates strongly due to the highly intermitent nature of the SF; on average, during the most stable phases, $t_{\rm d} \sim 20$ Myr. This time is a sort of average for the entire flow. The global dissipation rate is dominated by the dissipation at the injection zones (i.e., the former is very closely given by the latter times a "weight" which is essentially the filling factor of the injection zones), and $t_{\rm d}$ near these zones seems to be of the order of the crossing time $t_{\rm cross} = l_{\rm f}/u_{\rm f} \sim 1$ Myr. Although measuring $t_{\rm d}$ in the injections zones is difficult, AV00 test whether in the injection zones $t_{\rm d}$ is indeed proportional to $l_{\rm f}/u_{\rm f}$ by monitoring the global dissipation time.

Two additional runs where $l_{\rm f}$ and $u_{\rm f}$ are varied, were considered. In the former case, a value of $l_{\rm f}$ 4 times larger than in the fiducial run was used, while in the latter, a value of $u_{\rm f}$ 2 times smaller was used. In both runs, the global kinetic energy $E_{\rm k}$ content decreased roughly by a factor 2 with respect to the fiducial run. Moreover, the dissipation rate is expected to depend on $E_{\rm k}$, as $\dot{E}_k^{\rm diss} \propto E_{\rm k}^{3/2}$, as is the case for incompressible flows (see, e.g., Frisch[15]). This implies $t_{\rm d} \propto E_{\rm k}^{-1/2}$. This dependence was confirmed running a simulation identical to the fiducial one except for a decreased threshold density for SF which produces an increased SF in the simulation, and consequently the kinetic energy content. Now, after taking into account this dependence of t_d on E_k, the global dissipation times measured in the two runs mentioned above are still larger than in the fiducial run by factors of roughly 4 and 2, confirming that $t_\mathrm{d} \propto t_\mathrm{cross} \equiv l_\mathrm{f}/u_\mathrm{f}$. In conclusion, the following empirical relation for the global dissipation time was found:

$$t_\mathrm{d} \approx 1.8 \times 10^7 \mathrm{yr} \left(\frac{l_\mathrm{f}/30 \text{ pc}}{u_\mathrm{f}/30 \text{ km s}^{-1}} \right) \left(\frac{u_\mathrm{rms}}{6.6 \text{ km s}^{-1}} \right)^{-1}. \tag{1}$$

4 Dissipation in decaying turbulence

Due to the small-scale, scattered and short-duration nature of the energy input sources, most of the volume actually is occupied by a turbulent flow in a decaying regime. In order to study the decaying regime, AV00 considered a continuation run to the fiducial one, but turning SF off (see Fig. 2). Since in the decaying case the Coriolis force tends to form vortical structures which have extremely low dissipation rates, this force was not considered, as we feel it produces unrealistic situations. It was found that E_k decays as $(1 + t)^{-n}$ with $n \sim 0.8$, in reasonable agreement with previous studies for isothermal fluids (Mac Low $et\ al.$[8], Stone $et\ al.$[9]). A characteristic decay time, t_dec, can be defined as the time the initial E_k decays by a factor of 2. For $n \approx 0.8$, $t_\mathrm{dec} \approx 1.8 \times 10^7$ years, which is in agreement with t_d in the driven turbulence. With these timescales, "residual" turbulent motions propagating at roughly 10 km/s would attain typical distances of approximately 200 pc. AV00 also suggested, on the basis of dimensional arguments, that E_k and v_rms will decay with distance ℓ as ℓ^{-2m} and ℓ^{-m}, respectively, with $m = n/(2-n)$. For $n \approx 1$ this implies that E_k and u_rms decay with distance as ℓ^{-2} and ℓ^{-1}, respectively. Thus, "residual" turbulent motions with an rms velocity of roughly 6–10 km/s, as observed in our and other disk galaxies, would decay to roughly 1/10 of that value on distances $\sim$0.6–1 kpc, in the absence of energy sources.

5 Conclusions and implications

Using 2D simulations of MHD, compressible, self-gravitating flows in the presence of parametrized heating and cooling and nonubiquituos turbulent energy injection, we have found that:

• The spatially scattered, small-scale nature of the input sources in our ISM-like simulations gives rise to the coexistence of both forced and decaying turbulent regimes within the same flow.

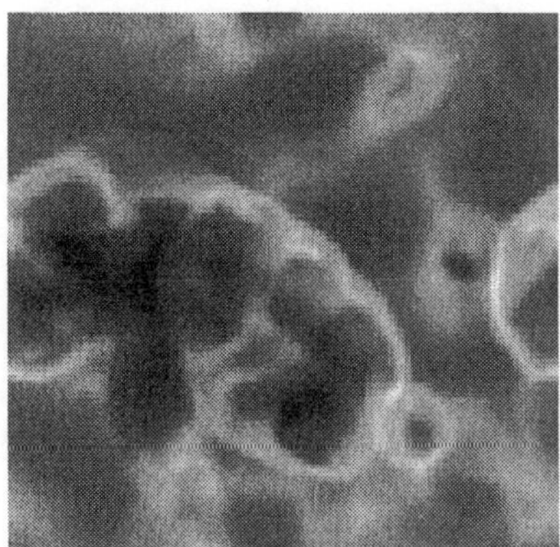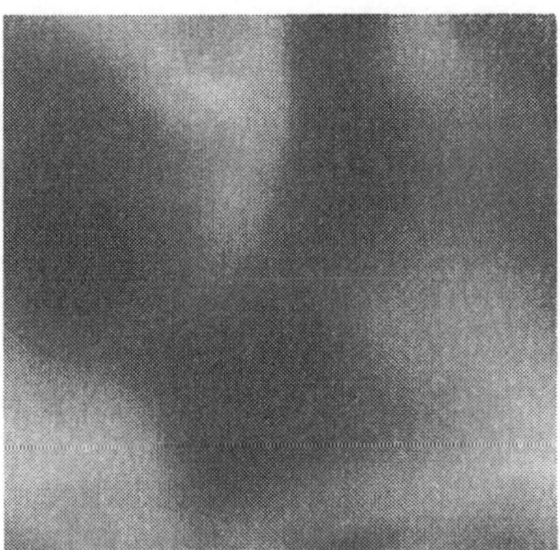

Figure 2. Log of the density field for the decay simulation at times $t = 0$ (*left*) and $t = 180.7$ Myr (*right*). This run is a restart of the fiducial run in forced regime at $t = 45.5$ Myr, with SF, Coriolis and self-gravity turned off. Expanding shells can still be seen at the earlier time, while a much smoother density structure is seen at the later time.

- The kinetic energy injection and dissipation rates are always very close to each other, suggesting that most of the dissipation occurs at or near the input sources.

- In the forced regime the *global* dissipation time $t_d \sim (l_f/u_f)(1/u_{rms})$; for $l_f \approx 30$ pc, $u_f \approx 30$ km/s, and $u_{rms} \approx 7$ km/s, we obtained $t_d \approx 2 \times 10^7$ years. This time is an average for the entire flow; near the injection zones the dissipation time is much shorter, of the order of the crossing time $t_{cross} = l_f/u_f$.

- In the decaying regime (far from the input sources), the "residual" turbulence decays as $E_k(t) \propto (1 + t)^n$ with $n \approx -0.8$. The characteristic decay time is again $\sim 2 \times 10^7$ years. From dimensional arguments, we suggest that E_k and u_{rms} of the "residual" turbulence should decay with distance ℓ as $\ell^{-2n/(2-n)}$ and $\ell^{-n/(2-n)}$, respectively. For $n \sim 1$, $E_k \sim \ell^{-2}$ and $u_{rms} \sim \ell^{-1}$.

If our results are applicable to the stratified vertical direction in the Galactic disk, turbulent motions produced near the disk plane will propagate up to distances of the order of the neutral gas disk half thickness. This is consistent with models of galaxy evolution where the H I disk thickness is supposed to be determined mainly by the turbulent kinetic energy content of the medium, which in turn it results from the balance between kinetic energy injection by stars and its turbulent dissipation rate (e.g., Scalo and Struck-Marcell[16], Vázquez and Scalo[17], Dopita[18], Firmani and Tutukov[19,20], Dopita and Ryder[21], Wang and Silk[22], Firmani *et al.*[23]). The dissipation timescales used in these models of galaxy evolution are of the order of a few 10^7 years for Milky Way-like galaxies; actually, they are 1.5–3 times larger than those we have found. Since in these models of self-regulating SF, which succesfully predict the observable H I disk thickness, the latter depends on t_d, our values of t_d could result in thinner H I disks than observed. However, in practice, the disk height may be sustained by factors other than turbulent pressure alone, such as the magnetic pressure and the cosmic ray pressure. Additionally, the vertical Galactic disk is stratified; this may facilitate the survival of turbulence to high Galactic latitudes. Finally, it is probable that some SF occurs far from the galactic midplane increasing the probability of energy escaping from the disk. Thus, we consider that the disk thicknesses estimated with our turbulent dissipation times are a lower limit. A study of turbulent dissipation in a stratified medium (vertical direction) is underway.

On the other hand, our results may pose a serious difficulty for models of galaxy formation where the turbulent kinetic energy injected by SNe and OB stars is assumed to reheat and drive back the gas from the disk into the cosmological dark matter halo (of sizes 15–20 times the disk size) in such a way that SF is self-regulated at the level of the whole intrahalo medium. In order for the turbulent gas to be driven back into the intrahalo medium, the typical sizes of the supershells (input sources) should exceed the gaseous disk height, in order to avoid the rapid dissipation in the disk. This could be a common situation in dwarf starbust galaxies (but see the results of hydrodynamical simulations by Mac Low and Ferrara[24]).

In conclusion, the results of our simulations imply that ISM turbulence is essentially different from ideal homogeneous turbulence due to the nature of the intervening energy injection (forcing) mechanisms. The intermittent, nonubiquitous, small-scale nature of interstellar kinetic energy injection seems to be responsible for a new, rich kind of turbulent flow, whose dissipation properties seem to pose clear constraints on models of galaxy formation and evolution.

Acknowledgements

The simulations were performed on the CRAY Y-MP 4/64 at DGSCA-UNAM. This work has received partial financial support from CONACYT grant 27752-E to E.V.-S.

References

1. C. F. McKee and J. P. Ostriker, *ApJ* **218**, 148 (1977).
2. A. Boulares and D. P. Cox, *ApJ* **365**, 544 (1990).

3. C. F. McKee, in *Evolution of the Interstellar Medium, PASP Conf. Series*, L. Blitz, ed., 3 (San Francisco: ASP, 1990).

4. F. J. Lockman and C. S. Gehman, *ApJ* **382**, 182 (1991).

5. A. Ferrara, *ApJ* **407**, 157 (1993).

6. C. A. Norman and A. Ferrara, *ApJ* **467**, 280 (1996).

7. V. Avila-Reese and E. Vázquez-Semadeni, *ApJ* submitted (2000).

8. M.-M. Mac Low, R. S. Klessen and A. Burkert, *Phys. Rev. Lett.* **80**, 2754 (1998).

9. J. M. Stone, E. O. Ostriker and C. F. Gammie, *ApJ* **508**, L99 (1998).

10. P. Padoan and A. Nordlund, *ApJ* **526**, 279 (1999).

11. M.-M. Mac Low, *ApJ* **524**, 169 (1999).

12. E. Vázquez-Semadeni, T. Passot and A. Pouquet, *ApJ* **441**, 702 (1995).

13. E. Vázquez-Semadeni, T. Passot and A. Pouquet, *ApJ* **473**, 881 (1996).

14. T. Passot, E. Vázquez-Semadeni and A. Pouquet, *ApJ* **455**, 533 (1995).

15. U. Frisch, *Turbulence* (Cambridge Univ. Press, Cambridge, 1995)

16. J. M. Scalo and C. Struck-Marcell, *ApJ* **276**, 60 (1984).

17. E. C. Vázquez and J. M. Scalo, *ApJ* **343**, 644 (1989).

18. M. A. Dopita, in *The Interstellar Medium in Galaxies*, H. A. Thronson, Jr. and J. M. Shull, eds., 437 (Dordrecht, Kluwer, 1990).

19. C. Firmani and A. V. Tutukov, *A&A* **264**, 37 (1992).

20. C. Firmani and A. V. Tutukov, *A&A* **288**, 713 (1994).

21. M. A. Dopita and S. D. Ryder, *ApJ* **430**, 163 (1994).

22. B. Wang and J. Silk, *ApJ* **427**, 759 (1994).

23. C. Firmani, X. Hernández and J. Gallagher, *A&A* **308**, 403 (1996).

24. M.-M. Mac Low and A. Ferrara, *ApJ* **513**, 142 (1999).

THE DENSITY PROBABILITY DISTRIBUTION FUNCTION IN TURBULENT, ISOTHERMAL, MAGNETIZED FLOWS IN A SLAB GEOMETRY

ENRIQUE VÁZQUEZ-SEMADENI

Instituto de Astronomía, Universidad Nacional Autónoma de México,
Apdo. Postal 70-264, 04510 México, D.F., México
E-mail: enro@astroscu.unam.mx

THIERRY PASSOT

Observatoire de la Côte d'Azur, BP 4229, Nice Cédex 4, FRANCE
E-mail: passot@obs-nice.fr

We investigate the behavior of the magnetic pressure, b^2, in fully turbulent MHD flows in "1+2/3" dimensions by means of its effect on the probability density function (PDF) of the density field. We start by reviewing our previous results for general polytropic flows, according to which the value of the polytropic exponent determines the functional shape of the PDF. A lognormal density PDF appears in the isothermal ($\gamma = 1$) case, but a power-law tail at either large or small densities appears for large Mach numbers when $\gamma > 1$ and $\gamma < 1$, respectively. In the isothermal magnetic case, the relevant parameter is the field fluctuation amplitude, $\delta B/B$. A lognormal PDF still appears for small field fluctuations (generally the case for *large mean fields*), but a significant low-density excess appears at large fluctuation amplitudes (*weak mean fields*), similar to the behavior at $\gamma > 1$ of polytropic flows. We interpret these results in terms of simple nonlinear MHD waves, for which the magnetic pressure behaves linearly with the density in the case of the slow mode, and quadratically in the case of the fast wave. Finally, we discuss some implications of these results, in particular the fact that the effect of the magnetic field in modifying the PDF is strongest when the mean field is weak.

1　Introduction

A fundamental feature of compressible turbulence is the formation of density fluctuations, a property of central interest in astrophysics, as density excesses ("clouds") in the turbulent interstellar medium (ISM) exhibit numerous statistical properties over a range of sizes[1,2] whose origin is not yet well understood. Although a full understanding of star formation requires knowledge of the full (multiple-point) statistics in order to determine mean densities as a function of region size, a first step towards this goal is the description and physical understanding of the one-point statistics, or *probability density function* (PDF) of the mass density field.

Previous studies have shown that the density PDF in turbulent compressible flows has a lognormal form in the isothermal case,[3-6] but exhibits a power-law tail at densities larger (smaller) than the mean for flows with polytropic exponents smaller (larger) than unity[7-9]. It should be noted that some of those works referred to purely hydrodynamic flows,[3,8] while the rest referred to magnetohydrodynamic (MHD) flows.

In particular, ref. [8] presented a heuristic model for the development of the lognormal and power-law PDFs, based essentially on the behavior of the speed of sound with density in those types of flows. In this sense, the density PDF is a diagnostic for the dependence of the pressure with density. However, such model applied only to nonmagnetic flows, while the real ISM is most likely magnetized to a significant extent. In this paper we present preliminary results on the density PDF of turbulent magnetized flows in "1+2/3" dimensions (i.e., variability is considered only with respect to the spatial variable x for the three components of the velocity and magnetic fields), as a first attempt to characterize the behavior of magnetic pressure with density in the fully turbulent regime. Previous works have focused on the pressure produced by weakly nonlinear Alfvén waves[10,11], but here we consider fully turbulent regimes with arbitrarily large magnetic fluctuation amplitudes.

2 The nonmagnetic case

The form of the density PDF in nonmagnetic flows and its relation to the effective equation of state of the system has been understood in terms of a heuristic model by Passot and Vázquez-Semadeni,[8] herafter PVS98 (see also ref. [9]). As in ref. [3], this model idealizes the generation of turbulent density fluctuations as a "multistep" process, in which the local density at any given point in the flow is the result of a series of jumps due to the continuous passage of shock waves. The generation of densities is an iterative multiplicative process, in which every new density is obtained through a jump from the previous one, giving for the final density ρ_f in terms of the intermediate steps ρ_i: $\rho_f = \rho_0 \prod_i (\rho_{i+1}/\rho_i)$. In the isothermal case, the density jump is given by M^2, where M is the Mach number of the shock and, because the speed of sound is constant, a given *velocity* jump is always characterized by the same Mach number, regardless of the local density, so that individual jumps can be regarded as independent, but extracted from the same distribution. In terms of the variable $s = \ln \rho$, the iterative process is *additive* so that the Central Limit Theorem can be applied in the limit of a large number of jumps, leading to a normal distribution for s and thus a lognormal distribution for ρ, as observed in the numerical experiments mentioned above.

Concerning the variance of the distribution, PVS98 have suggested (and confirmed numerically), through an analysis of the shock and expansion waves in the system, that for a large range of Mach numbers the typical size of the logarithmic jump is expected to be $\sigma_s \sim M_{\mathrm{rms}}$, where M_{rms} is the rms Mach number. The mean s_0 of the distribution can be directly evaluated from the mass conservation condition $\langle \rho \rangle = \int_{-\infty}^{+\infty} e^s P(s) ds = 1$, where $P(s)$ is the PDF of s, yielding $s_0 = -\sigma_s^2/2$. The isothermal model PDF for s thus reads

$$P(s)ds = \frac{1}{\sqrt{2\pi\sigma_s^2}} \exp\left[-\frac{(s-s_0)^2}{2\sigma_s^2}\right] ds, \tag{1}$$

with $\sigma_s^2 = \beta M_{\mathrm{rms}}^2$ and β a proportionality constant of order unity.

In the general polytropic case where the pressure P behaves as $P \propto \rho^\gamma$, with γ the polytropic exponent, the local Mach number $M(s)$ at a density $\rho = e^s$ is related to the one at the mean density M by $M(s) = M \exp^{(1-\gamma)s/2}$, suggesting an ansatz where the PDF keeps the same dependence on M, provided the above replacement is made. After relocating the term in s_0 from inside the exponential function to the normalization constant, the model PDF for the polytropic case reads

$$P(s;\gamma)ds = C(\gamma) \exp\left[\frac{-s^2 e^{(\gamma-1)s}}{2M^2} - \alpha(\gamma)s\right] ds. \tag{2}$$

This equation shows that when $(\gamma-1)s < 0$, the PDF of ρ asymptotically approaches a power law, while in the opposite case it decays faster than a lognormal. Thus, for $0 < \gamma < 1$, the PDF approaches a power law at large densities $(s > 0)$, and at low densities $(s < 0)$ for $\gamma > 1$ (Fig. 1).

3   The magnetic case

In what follows we restrict ourselves to the isothermal case $(\gamma = 1)$ and to a propagation along a uniform ambient magnetic field B_0 and concentrate on the deviations from the corresponding lognormal PDF induced by the magnetic field.

3.1   Numerical results

The numerical simulations solve the MHD equations using a pseudospectral method in a slab geometry ("1+2/3D"). A resolution of 2048 grid points is used, which allows the handling of large enough Mach and Reynolds numbers with only regular second-order viscosity. A random acceleration is applied on the y and z components of the velocity field, i.e., transverally to B_0, thus inducing Alfvén waves into the flow. Since these waves have finite amplitudes, they in turn induce fast and slow magnetosonic waves. The simulations are

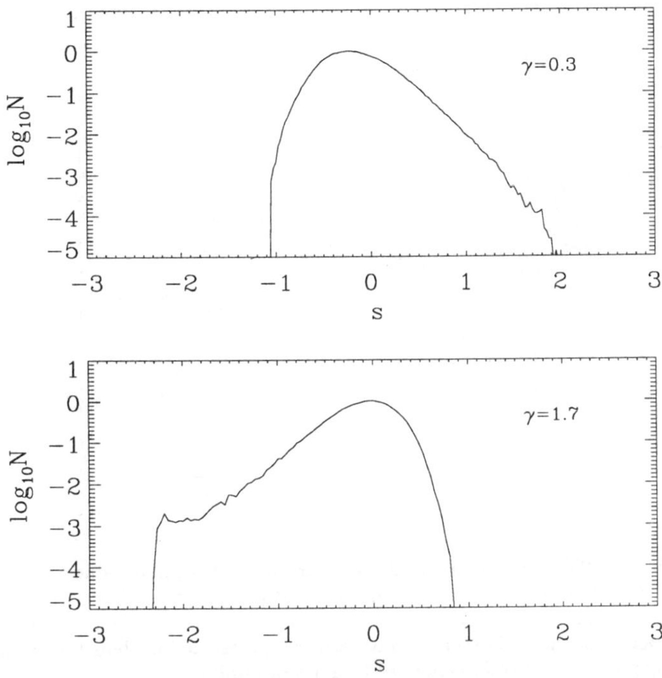

Figure 1. Density PDFs for two 1D simulations of polytropic turbulence, one with $\gamma = 0.3$ (left), and the other with $\gamma = 1.7$ (right), both with $M = 3$.

evolved over long times (several tens to a few hundred crossing times at the rms flow velocity) in order to obtain meaningful statistics for the density PDFs.

The simulations are essentially characterized by the sonic and Alfvénic Mach numbers, respectively defined as $M_s \equiv u/c_s$ and $M_a \equiv u/v_A$, where u is the rms velocity, c_s the sound speed and $v_A \equiv B_0^2/\rho$ the Alfvén speed. In order to investigate the effect of the magnetic field exclusively, we consider two runs (denoted I and II) with approximatively the same value of M_s (4.06 and 3.74 respectively) but very different values of M_a (0.36 and 1.66 resp.). These quantities are integrated over the duration of the run, excluding the first few time units in order to avoid including the uniform-density initial conditions. The rms relative field fluctuations $\delta B/B$ are for runs I and II respectively 0.75 and 6.65.

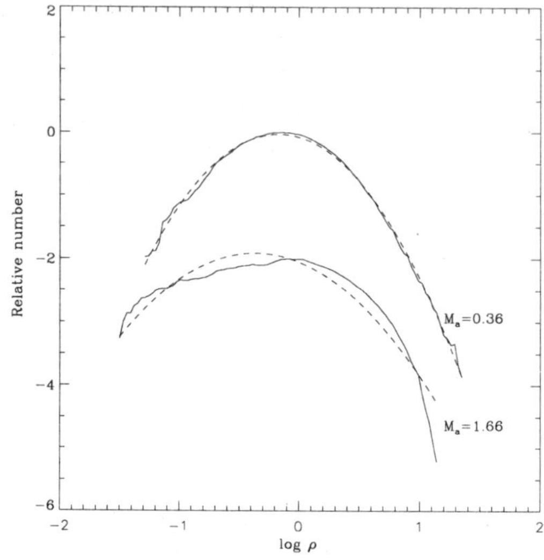

Figure 2. Density PDFs of two magnetic simulations with similar values of the sonic Mach number M_s but different values of the Alfvénic Mach number M_a. The simulation with small M_a (large mean field) is seen to have a nearly lognormal PDF, while the one with large M_a (weak mean field) is seen to have a large excess at low densities, indicative of an effective "magnetic" polytropic exponent larger than unity.

In Fig. 2 we show the PDFs for the two magnetic runs, together with lognormal fits (dashed lines). It is clearly seen that, contrary to what one might expect, the high-M_a run, which should be closest to the pure-hydro case, exhibits a strong deviation from lognormality, while the low-M_a PDF is quite close to a lognormal. This implies that the case with the weakest mean field is the one with the strongest effect of magnetic pressure on the density PDF.

3.2 Interpretation in terms of simple MHD waves

A preliminary interpretation of the behavior reported in the previous section can be given in terms of the so-called "simple MHD waves" (see, e.g., ref. [12]). These are the finite-amplitude equivalent of linear MHD waves, and have the same three well-known modes: Alfvén, slow and fast. A simple wave refers to a solution that depends only on a single variable, any combinations of the spatial (x) and temporal (t) independent variables. However, as mentioned above, simple waves have finite amplitudes, and in general they develop shocks in a finite time.

A number of properties of simple waves has been reported by Mann[12] (see also Jeffrey and Taniuti[13]). Among them, most relevant for the present discussion are the propagation velocities of the three modes, and the relations between the density and magnetic field fluctuations. The Alfvén mode with speed V such that $V^2 = v_{\mathrm{A}x}^2$ is not associated with density fluctuations. The fast and slow speed are given by

$$V_\pm^2 = \frac{v_{\mathrm{A}}^2 + c_{\mathrm{s}}^2}{2}\left\{1 \pm \left[1 - \frac{4c_{\mathrm{s}}^2 v_{\mathrm{A}x}^2}{(v_{\mathrm{A}}^2 + c_{\mathrm{s}}^2)^2}\right]^{1/2}\right\} \tag{3}$$

and the relation between the magnitude of the magnetic field b and the density is given by

$$\frac{db}{d\rho} = \frac{V^2 - c_{\mathrm{s}}^2}{b}, \tag{4}$$

where $v_{\mathrm{A}x}^2 = b_x^2/\rho$, $v_{\mathrm{A}}^2 = b^2/\rho$, and c_{s} is the sound speed. An important point to note is that Eqs. (3) and (4) imply a positive correlation between b and ρ for the fast mode (V_+) and an anticorrelation for the slow mode (V_-),[12].

Inserting Eq. (3) in Eq. (4), one can find the density dependence of the magnetic pressure. Excluding the case where the β of the plasma is of order unity with at the same time small to moderate field distorsions, this dependence simplifies and reads (general numerical solutions have been presented by Mann[12]):

$$\frac{b^2}{b_x^2} \approx a_1 - a_2 \frac{c_{\mathrm{s}}^2 \rho}{b_x^2} \qquad \text{(slow mode)} \tag{5}$$

$$b^2 \approx 2a_3\rho^2 \quad \text{(fast mode)}, \tag{6}$$

where $b^2 = b_x^2 + b_\perp^2$, with $b_\perp^2 = b_y^2 + b_z^2$ being the magnitude of the magnetic field fluctuation. Note that in our 1+2/3 geometry, $b_x\,(= B_0)$ is constant. The quantities a_i denote integration constants.

From these relations, it can be seen that the magnetic pressure, $\propto b^2$, scales roughly linearly with the density (albeit inversely) in the case of the slow mode, but quadratically in the case of the fast mode. Note also that, in the case of small field fluctuations (most often the case if $\beta \equiv c_{\mathrm{s}}^2/(b_x^2/\rho) \ll 1$) the coefficient of ρ in Eq. (5) is small, implying that, for the slow mode, large density fluctuations can occur even for small variations of b. Thus, *the slow mode is expected to dominate density fluctuation production in the case of small field fluctuations.* Instead, at large enough field fluctuations, the quadratic dependence of b^2 on ρ of the fast mode eventually overwhelms the linear dependence of the slow mode (and also of the thermal pressure), so *the fast mode is expected to dominate the density fluctuation production.*

The dependence of the magnetic pressure on the density seems to describe the observations of Sec. 3.1 adequately because, according to the results of the nonmagnetic case, a pressure that depends linearly on the density produces a lognormal density PDF, while one with a higher-than-linear dependence produces a near power law at low densities.

4   Conclusions

4.1   Summary

In this paper we have reviewed our previous results[8] on the development of the density PDF as a consequence of the effective (polytropic) equation of state, and applied them to an understanding of the PDF in the magnetic case.

In the context of the model presented in ref. [8], isothermal nonmagnetic flows have lognormal mass density PDFs while in the nonisothermal cases, a power-law tail develops at high Mach number, at either $\rho > \langle \rho \rangle$ or $\rho < \langle \rho \rangle$ depending on whether $\gamma < 1$ or $\gamma > 1$, respectively.

In the magnetic case, we considered only isothermal cases and a propagation parallel to the ambient magnetic field, but reported that a deviation from the lognormal PDF occurs *when the magnetic fluctuations are large*, a case expected when the mean field is small. We interpreted this effect in terms of so-called "simple", finite-amplitude nonlinear MHD waves. Indeed, for the slow mode of these waves, the magnetic pressure b^2 behaves linearly with the density ρ, while for the fast mode, the magnetic pressure behaves quadratically with ρ. Together with the fact that for the slow wave the density is weighted by a small factor, this suggests that the production of density fluctuations is dominated by the slow waves at small field deviations (i.e., B_0 large) giving a lognormal PDF again, while for weak mean fields (large field deviations) the quadratic density dependence of b^2 produces an excess at small densities in the PDF, corresponding to the $\gamma > 1$ case of the polytopic description. A more detailed study, including in particular the case of nonparallel propagation is in progress.

4.2   Astrophysical implications

The results of this paper have a number of implications in the astrophysical context. First, we have seen that the main parameter determining the effect of the magnetic field appears to be the magnetic fluctuation amplitude, $\delta B/B$, rather than β alone, which is the parameter most frequently used to characterize MHD flows. In other words, it is important to have a knowledge of the relative importance of the turbulent fluctuations (as given by the Alfvénic Mach number) in addition to simply the ratio of thermal to magnetic pressures (as given by β).

Second, as far as the density PDF is concerned, the effect of the magnetic field is most notorious when the field fluctuations are large (generally, when the mean field is weak). This suggests that the limit of vanishing field does not approach the nonmagnetic case, and in this sense the latter is singular. This may imply that it is inadequate to model the ISM as a nonmagnetic flow even if the degree of magnetization is very low, as has started to be recently contended at the level of molecular clouds (see, e.g., ref. [14]).

Acknowledgements

Part of this work was completed while the authors enjoyed the hospitality of the Institute for Theoretical Physics at the University of California at Santa Barbara, as participants of the Astrophysical Turbulence Program. This work has received partial funding from grants Conacyt (México) 27752-E to E.V.-S, the CNRS National Program "PCMI" to T.P. and NSF PHY94-07194.

References

1. R. B. Larson, *MNRAS* **194**, 809 (1981).
2. L. Blitz, in *Protostars and Planets III*, E. Levy and J. Lunine, eds. (Univ. Arizona, Tucson, 1993).
3. E. Vázquez-Semadeni, *ApJ* **423**, 681 (1994).
4. P. Padoan, Å. Nordlund and B. J. T. Jones, *MNRAS* **288**, 145 (1997).
5. E. C. Ostriker, C. F. Gammie and J. M. Stone, *ApJ* **513**, 259 (1999).
6. E. C. Ostriker, J. M. Stone and C. F. Gammie, ApJ, in press.
7. J. Scalo et al, *ApJ* **504**, 835 (1998).
8. T. Passot and E. Vázquez-Semadeni, *Phys. Rev. E* **58**, 4501 (1998).
9. Å. Nordlund and P. Padoan, in *Interstellar Turbulence*, J. Franco and A. Carraminñana, eds. (Cambridge Univ., Cambridge, 1999).
10. C. F. McKee and E. G. Zweibel, *ApJ* **440**, 686 (1995).
11. S. Spangler, *Phys. Fluids* **30**, 1104 (1987).
12. G. Mann, *J. Plasma Phys.* **53**, 109 (1995).
13. A. Jeffrey and T. Taniuti, *Non-linear Wave Propagation, with Applications to Physics and Magnetohydrodynamics* (Academic Press, New York, 1964).
14. P. Padoan and Å. Nordlund, *ApJ* **526**, 279 (1999).

MULTIFRACTAL STRUCTURE IN SIMULATIONS AND OBSERVATIONS OF THE INTERSTELLAR MEDIUM

WILDER CHICANA NUNCEBAY AND ENRIQUE VÁZQUEZ-SEMADENI

Instituto de Astronomía, Universidad Nacional Autónoma de México,
Apdo. Postal 70-264, 04510 México, D.F., México
E-mail: wilder,enro@astroscu.unam.mx

Multifractal analysis is a powerful technique that allows categorization of the structure of complex objects. We present preliminary results on the multifractal spectrum $f(\alpha)$ of observational and numerically simulated interstellar medium (ISM) data. We consider numerical simulations in two and three dimensions of the ISM at intermediate scales (hundreds of pc), as well as molecular gas data. For purely fractal objects, the multifractal spectrum is reduced to a single point. When applied to observational and simulated ISM data, the technique shows well-defined multifractal spectra in all cases, including velocity-channel projections of 3D density cubes, implying that the ISM has a multifractal structure, rather than being a simple fractal. For density data, *the quantity α corresponds to the scaling exponent between mass and size.* The fact that a range of values of α is found in all cases supports previous claims (Vázquez-Semadeni, Ballesteros-Paredes and Rodríguez 1997) that this exponent is not unique. Additionally, for density data, a significant fraction of the $f(\alpha)$ curve lies on values of α smaller than the dimension of the embedding space, indicating hierarchical structuring of the field (larger structures have smaller average densities). This property is also satisfied by projected velocity-channel data. Three- and two-dimensional simulations show significantly different $f(\alpha)$ curves, which suggests intrinsically different geometrical properties.

1 Introduction

Multifractals are systems in which a "measure" (e.g., the mass density) is distributed in an embedding space (called the "support") in such a manner that its scaling with size, and the region of space over which the measure scales with a particular value of the exponent, are both fractals. In recent years, the fractal structure of molecular clouds has been studied (e.g., Scalo[1], Falgarone *et al.*[2], Stutzki *et al.*[3]) and debated (e.g., Blitz and Williams[4]), but, given the fluid, turbulent nature of the interstellar medium (ISM), it is likely that it is rather a multifractal (e.g., Sreenivasan[5], Chappell and Scalo[6]) at least within a certain range of scales. In this work, we report preliminary results from the application of multifractal analysis (MFA) to both numerical simulations and observational data of the ISM.

It should be emphasized that MFA does *not* involve any definition of "clouds", thus having the advantage of being free of biases due to selection criteria, but the disadvantage of not being directly comparable with studies based on cloud surveys.

2 Definitions and method

We use the so-called "moment" method for computing the multifractal spectrum $f(\alpha)$, where α is the scaling exponent of the "measure"[a] with size, and $f(\alpha)$ is the fractal dimension of the subset of space in which the measure scales with exponent α. Strictly speaking, α is a pointwise dimension, and $f(\alpha)$ is similar to a box-counting dimension (see, e.g., Ott[7]).

In the moment method, the image is partitioned into boxes of size ϵ. If $\mu_i(\epsilon)$ is the measure integrated over box i, then α and $f(\alpha)$ are given by

$$\alpha(q) = \lim_{\epsilon \to 0} \left[\frac{\sum_{i=1}^{N(\epsilon)} \Gamma_i(q) \log \mu_i}{\log(\epsilon)} \right] \tag{1}$$

$$f(\alpha(q)) = \lim_{\epsilon \to 0} \left[\frac{\sum_{i=1}^{N(\epsilon)} \Gamma_i(q) \log \Gamma_i(q)}{\log(\epsilon)} \right], \tag{2}$$

where

$$\Gamma_i(q) = \frac{\mu_i^q(\epsilon)}{\sum_{j=1}^{N(\epsilon)} \mu_j^q} \tag{3}$$

are the qth moments of the measure, and the sums run over all boxes of size ϵ in the image. Operationally, we proceed in the usual way of dividing the data in boxes of different sizes and taking α and $f(\alpha)$ as the slopes of the scalings between the sums and the box size (e.g., Chhabra and Jensen[8]). We have tested our numerical algorithm on 1D, 2D and 3D binomial multifractals (constructed by sequentially subdividing a subset of space, assigning to each new subdivision a fraction of the measure contained in the parent division). We checked that the maximum of the $f(\alpha)$ curve coincides with the dimension of the geometric support (the space in which the multifractal is embedded), as expected in this case, and that the shape of the curve coincides with examples in the literature (e.g., DeGraff[9]).

---

[a] In this paper, the measure corresponds to the mass in 3D and 2D simulations and in projected images (velocity-integrated or single-channel maps) of 3D data, and to the velocity- and area-integrated emission in the ^{13}CO map.

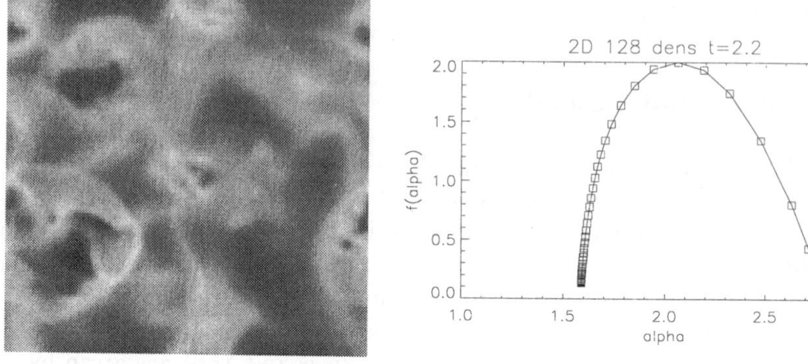

Figure 1. (a) (left) Logarithmic image of the density field of the 2D numerical simulation at time $t = 2.2$ (in arbitrary code units). (b) (right) $f(\alpha)$ spectrum for the density field of panel (a). Note the positive skewness of the curve.

3   The data

We consider two kinds of data sets, numerical and observational. The former are taken from 2- and 3-dimensional simulations of the turbulent ISM, which respectively represent boxes of sizes 1 kpc and 300 pc on the Galactic plane centered at the solar Galactocentric distance. The simulations include parameterized radiative heating and cooling, self-gravity, the magnetic field and the Galactic rotation, as in the model of Passot et al.[10]. The 3D simulation is taken from Pichardo et al.[11].

The observational data is a velocity-integrated ^{13}CO map of the Taurus region taken from Mizuno et al.[12].

4   Results

We have computed the $f(\alpha)$ spectra for the 2- and 3-dimensional density fields of the simulations, as well as for an integrated column-density map and a single-velocity column-density channel map (of width 1 km s^{-1}) of the 3D simulation. Images of these data are respectively shown in panels (a) of Figs. 1 through 5, while their $f(\alpha)$ spectra are shown in panels (b) of the same figures. The ^{13}CO velocity-integrated map and its corresponding $f(\alpha)$ curve are shown in Fig. 6. We have the following preliminary results.

1. All data sets give well-defined multifractal spectra, which indicates that they are indeed multifractals, rather than simple fractals. In fact, a simple

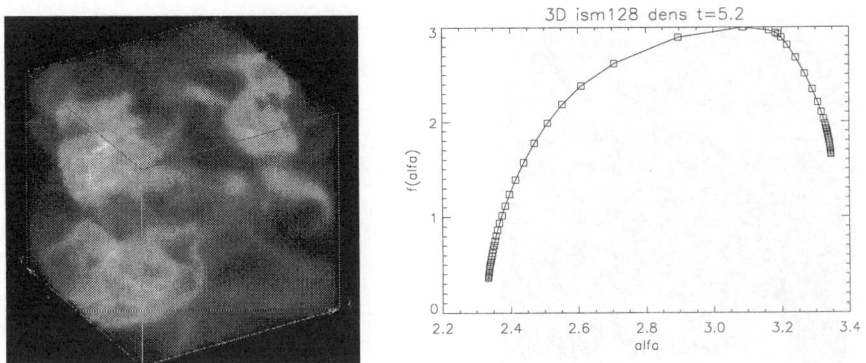

Figure 2. (a) (left) Density field of the 3D simulation at $t = 5.2$. (b) (right) $f(\alpha)$ curve for the density field of panel (a)). In this case the skewness is negative.

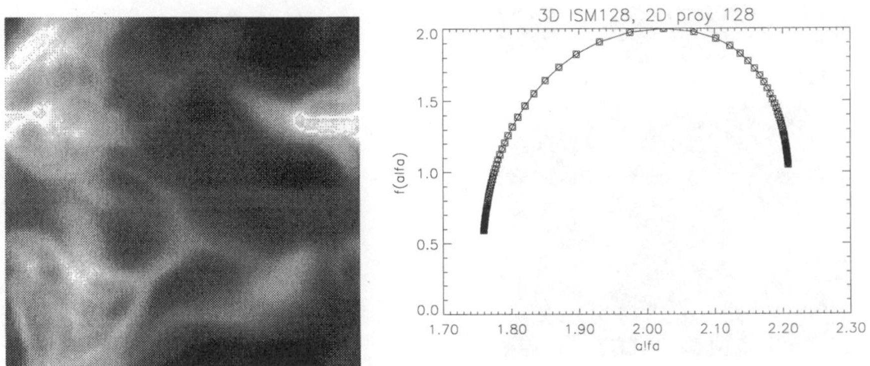

Figure 3. (a) (left) Total density projection along the z axis of the 3D simulation shown in Fig. 2. (b) (right) $f(\alpha)$ curve for the column density field of panel (a). The projection produces a less skewed $f(\alpha)$ curve, but still with negative skewness.

fractal refers to the geometry of a spatial object, but not to how another property, such as the mass density, is distributed within the object. This is what is naturally described by the multifractal spectrum. Thus, the multifractal description naturally befits the spatial distribution of gas density in the ISM.

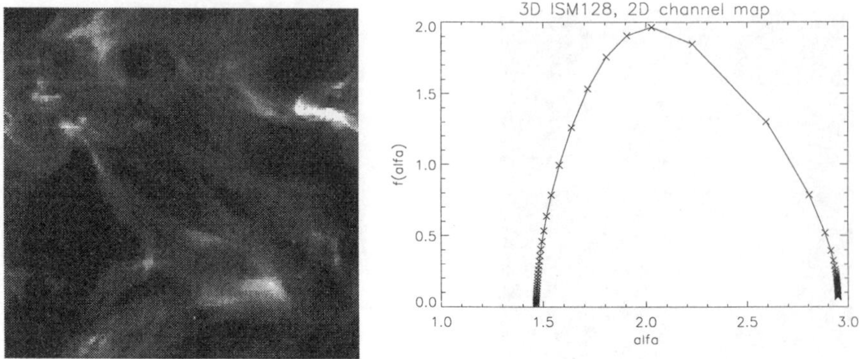

Figure 4. (a) (left) Column density map for the central channel of a 16-channel position–velocity cube for the run of Fig. 2. The velocity width for the channel is 0.75 km s^{-1}. (b) (right) $f(\alpha)$ curve. The skewness becomes positive.

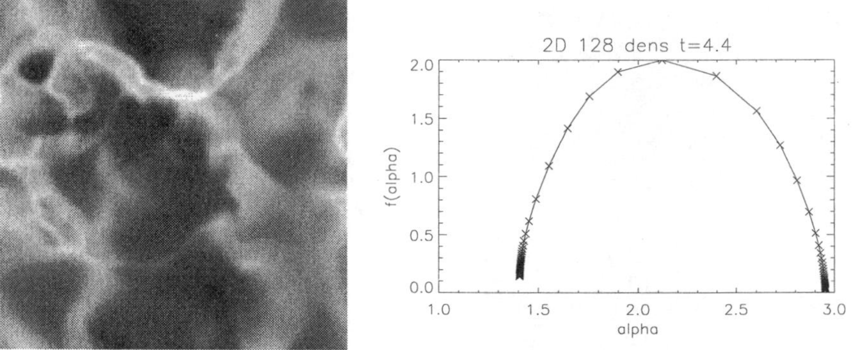

Figure 5. (a) (left) Logarithmic image of the density field of the 2D numerical simulation at time $t = 4.4$. (b) (right) $f(\alpha)$ spectrum. Compare to Fig. 1. The $f(\alpha)$ spectrum has become considerably wider.

2. The quantity α is the scaling exponent with size of the measure (either mass, total integrated mass along a line of sight (LOS) or LOS-integrated mass column density in the velocity channel map). In particular, if the measure is the mass, then the density scales with size as $\rho \propto R^{\alpha - D}$, where D is the dimension of the embedding space. It is seen in the figures that the range spanned by α includes a significant proportion of

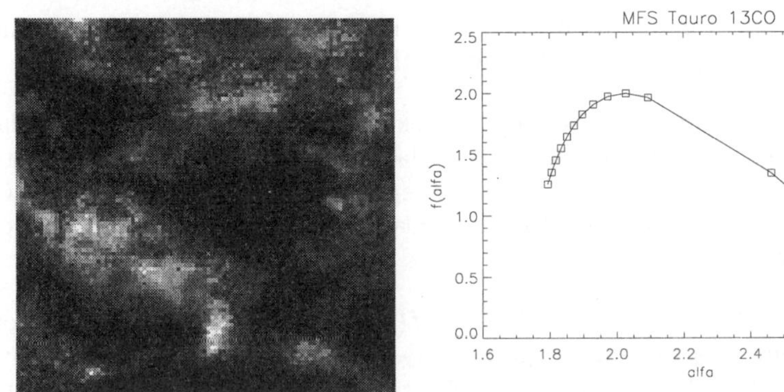

Figure 6. (a) (left) Velocity-integrated ^{13}CO map of the Taurus region. (b) (right) $f(\alpha)$ curve. In this case, the skewness is positive, contrary to the case of the integrated density field of the 3D simulation (Fig. 4).

values smaller than D, for which smaller structures have larger densities. This is consistent with a hierarchically nested structure of the medium, in which smaller, denser structures occur inside larger, less dense ones. However, there are also α values larger than D, since not all structures are nested within each other. It is possible that a larger structure may have a higher mean density than a smaller one, if the two are separated from each other.

3. The existence of a finite range of α values implies that there is not a unique scaling exponent for mass with size, contrary to what would be implied for example by Larson's density–size scaling relation[13]. However, there is not necessarily a conflict between our result and Larson's relation, since the latter is biased to high-column density clouds, while the multifractal spectrum refers to the whole density or column-density fields. This result is consistent with that of Vázquez-Semadeni *et al.*[14], which showed that in 2D simulations of the ISM, "clouds" (connected regions of density above a given threshold) do not define a single scaling density–size relation, but Larson's relation seems to be an upper bound to the region occupied by the clouds in a density–size diagram.

4. The $f(\alpha)$ curves are in general not symmetric. In particular, it is noteworthy that the $f(\alpha)$ curve for the 3D density field has a negative skewness, while that for the 2D density field has a positive skewness, which

suggests that fundamental differences in the spatial structuring of the density occur upon going from two to three dimensions.

5. From Figs. 2 and 6 we note that the skewness of the $f(\alpha)$ curve for the column–density map of the 3D simulation is reversed with respect to that of the LOS-integrated map of the Taurus region. At this time we do not know the reason for this discrepancy, which will be the subject of upcoming work.

6. From Figs. 2 and 5, we note that the $f(\alpha)$ curve for the same simulation at different times changes significantly, suggesting that there exists no universality in the compressible, self-gravitating MHD case, contrary to the situation for incompressible turbulence (see, e.g., Sreenivasan[5]).

5   Future work

We have presented here only preliminary results and interpretations of the multifractal spectra of the numerical and observational fields. We now plan to investigate the relationship between the intrinsic dynamics of the flow and both the range of α values and the shape of the $f(\alpha)$ curve. In particular, we plan to consider sets of simulations with increasing numbers of physical ingredients, in order to find out at which point the universality known for incompressible flows[5] is broken, and also to determine the geometrical features that determine the skewness of the $f(\alpha)$ curve, as well as their dynamical origin.

Acknowledgments

We wish to thank Dr. A. Mizuno for kindly providing the ^{13}CO data. The 3D simulation was performed on the CRAY C98 of IDRIS, France. The 2D simulations were performed on the CRAY Y-MP 4/64 at DGSCA, UNAM. This project has received financial support from CONACYT grant 27752-E to E. V.-S.

References

1. J. Scalo, in *Physical Processes in Fragmentation and Star Formation*, R. Capuzzo-Dolcetta, C. Chiosi and A. di Fazio, eds., 151 (Kluwer, Dordrecht, 1990)
2. E. Falgarone, T. G. Phillips and C. K. Walker, *ApJ* **378**, 186 (1991).

3. J. Stutzki, F. Bensch, A. Heithausen, V. Ossenkopf and M. Zielinski, *A&A* **336**, 697 (1998).

4. L. Blitz and J. P. Williams, *ApJ* **488**, 145L (1997).

5. K. R. Sreenivasan, *Ann. Rev. Fluid Mech.* **23**, 539 (1991).

6. D. Chappell and J. Scalo, *ApJ* in press (2000).

7. E. Ott, *Chaos in Dynamical Systems* (Cambridge University Press, Cambridge, 1993).

8. A. Chhabra and R. Jensen, *Phys. Rev. Lett.* **62**, 1327 (1989).

9. D. R. DeGraff, *Multifractal Analysis of Radio Lobes*, Ph.D. Thesis (The University of North Carolina, Chapel Hill, 1992)

10. T. Passot, E. Vázqucz-Semadeni and A. Pouquet, *ApJ* **455**, 536 (1995).

11. B. Pichardo, E. Vázquez-Semadeni, A. Gazol, T. Passot and J. Ballesteros-Paredes, *ApJ* **532**, 353 (2000).

12. A. Mizuno, T. Onishi, Y. Yonekura, T. Nagahama, H. Ogawa and Y. Fukui, *ApJ* **445**, L161 (1995).

13. R. B. Larson, *MNRAS* **194**, 809 (1981).

14. E. Vázquez-Semadeni, J. Ballesteros-Paredes and L. F. Rodríguez, *ApJ* **474**, 292 (1997).

PART II

GEOPHYSICS

STRATIFORM LOW CLOUDS: PHENOMENOLOGY AND LARGE EDDY SIMULATIONS

DOUGLAS K. LILLY

University of Oklahoma and National Severe Storms Laboratory

Stratiform low clouds have important effects as weather events, particularly for aircraft terminal operations, and as a major influence on regional and global climatology. The structure and modeling of such clouds is reviewed, with emphasis on their microphysics and radiative influences. Large eddy simulation has been applied to stratiform clouds with some success, though limited by resolution.

1 Motivations and data sources

Two principal purposes motivate the study of stratiform low clouds, their immediate weather effects and their cumulative effects on climate. Low clouds and fog, which may be low clouds intersecting an elevated surface, inhibit transportation, especially aircraft terminal operations. The major U.S. air terminal which is most frequently subject to instrument flight rules is San Francisco, which is notoriously covered by stratus and stratocumulus clouds in the summer. At San Francisco, the summer stratus usually clears between 10 and 11 local time, but that is during and after the morning operations peak. Short-term predictability of the clearing time has been shown to be generally poor (Hilliker and Frisch[1]).

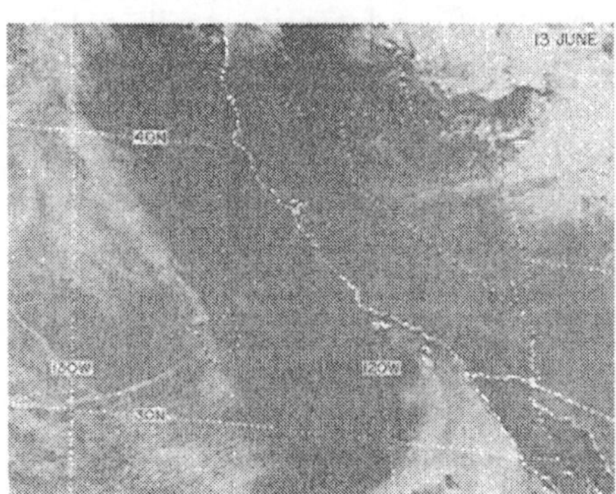

Figure 1. SMS/GOES satellite pictures for 1715 GMT on (left) 13 and (right) 17 June, 1976. From Brost, *et al.*[2]

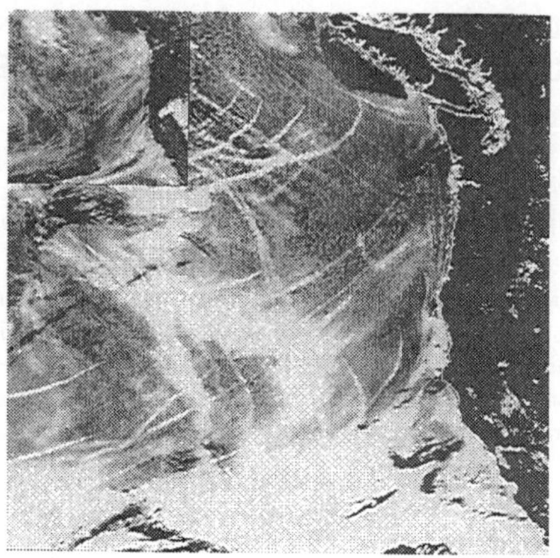

Figure 2. Satellite images showing an extensive stratocumulus cloud system off the West Coast of the U.S. The large image was constructed from data from a shortwave infrared wavelength, whereas the inset was constructed from reflected solar radiation measurements. The streaks on the infrared image are due to a reduced droplet size in cloud contaminated by the exhausts of ships. From King et al.[3]

Figure 1 shows satellite photographs of the eastern Pacific on summer days. On the right panel much of the California and Oregon coast and adjacent ocean areas are covered by a stratocumulus cloud layer, while on the left panel they are almost completely clear. This latter is an occasional event, the predictability of which is not very good. Figure 2 shows a remarkable discovery of satellite imagery, the appearance of condensation trails behind ships in the eastern Pacific. A study of this phenomenon is described in a later section.

An important long-term economic and social effect of stratiform cloudiness is its influence on the local and global climate. About 1/4th of the global ocean surface is covered at any time by low clouds. Provided they are not overlapped by deep higher clouds, they intercept, scatter, and partially reflect the incoming solar radiation, and therefore act as cooling elements in climate processes. They do not contribute much to greenhouse warming because their temperature is not much different from that of the surface beneath them. This is very different from the effect of high cloud layers, which partially block the outgoing radiation from the ground and radiate outward at a much colder temperature, thereby acting as heating elements. Figure 3 shows some nearly global maps of the frequency of low cloud cover over the oceans. The first panel is a map of satellite-observed low clouds in July, while the second shows

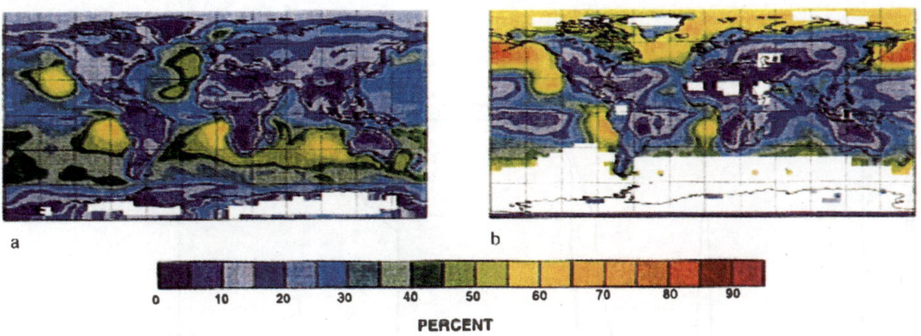

Figure 3. Distribution of low cloudiness (a) as observed by ISCCP (July 1983–1988 mean); (b) as observed from the surface by Warren *et al.*[4,5], showing Warren cloud atlas mean June–July–August cloud amount. From Randall *et al.*[6]

observations made from the surface, primarily from ships. The difference between these is the effect of overlapping higher clouds, which is frequent in the subpolar latitudes but much less so in the eastern ocean subtropic regions of frequent stratocumulus decks. As discussed later, direct human effect on the frequency of low clouds and their cooling effects is quite likely, which is a principal reason for the high emphasis given to low cloud layers.

Several field observation studies have been carried out to explore the structural details and evolution of low cloud layers. The largest and best known are the FIRE project off the California coast in 1987 (Albrecht *et al.*[7]), an extension into the eastern Atlantic, with the acronym ASTEX, in 1992 (Bretherton and Pincus[8], Bretherton *et al.*[9], Duynkerke *et al.*[10]), several programs in the North Sea and Atlantic near Britain (James[11], Slingo *et al.*[12], Nicholls and Leighton[13], Nicholls and Turton[14]), and the MAST ship track experiment off the California coast in 1994, reported by Durkee *et al.*[15]

2   Mixed layer theory

Cloud layers which are subject to buoyant forcing, either heated at the bottom or cooled at the top, are typically characterized as being well-mixed, that is the vertical gradients of the horizontal means of several fields are small, but have sharp interfaces at their tops. Fig. 4 shows two examples of cloud-topped mixed layers. The one with the higher cloud base appears to be well-mixed to the surface, while the other shows a layer of weak static stability in the lowest

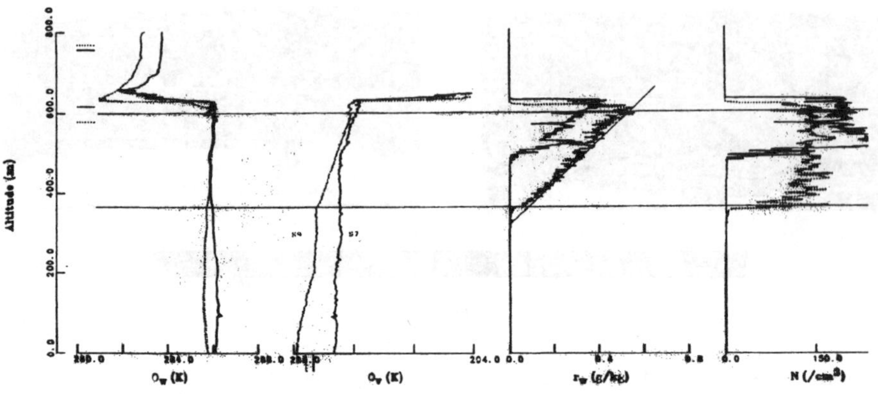

Figure 4. Aircraft soundings of wet-bulb potential temperature, θ_w, virtual potential temperature, θ_v, liquid water mixing ratio, r_w, and drop concentration, N, for two flights. The sloping line is the wet adiabatic liquid water slope. From Rogers and Telford.[16]

200 m of the virtual potential temperature panel, indicating a decoupling of the turbulence in the cloud layer from that in the lowest 200 m. This decoupling often results from effects of solar heating of the cloud layer.

Well-mixed boundary layers, with and without clouds, are subject to methods of fairly simple analysis, which are often useful for interpreting observed and simulated data (Lilly[17], Deardorff[18,19], Bretherton and Wyant[20]). The assumption of vertical uniformity of various quantities within the mixed layer allows the rate of change of the quantities to be determined from the fluxes at the top and bottom of the mixed layer. Typical quantities for which the mixed layer assumption may be suitable are liquid water potential temperature, total water mixing ratio, and passive scalars such as ozone mixing ratio. Surface turbulent fluxes are typically determined from a surface layer parameterization or may be arbitrarily specified. The turbulent flux at the top of the mixed layer can be shown[17] to equal the rate of entrainment of the property above the mixed layer

Determining that entrainment rate is the most difficult aspect of the theory. It is usually assumed to be related to the turbulent intensity at the top of the mixed layer, which is determined by the generation, dissipation, and turbulent transport of kinetic energy within the mixed layer. Attempts have been made to predictively solve the turbulent energy Equation, but that is a somewhat uncertain process, due to important higher order terms. Usually some fairly simple algorithm is Employed, relating the buoyancy flux at the top of the mixed layer to that at the bottom or to its vertical integral, and from the buoyancy flux the other relevant fluxes may be determined. Unfortunately, no such simple algorithm has been found that appears reliable under all circumstances. Partial reviews of the problem appear in Nicholls and Turton[14] and VanZanten *et al.*[21]

It is doubtful that mixed layer models can be effectively extended to cover the important cloud physics and chemistry issues to be discussed next. This is a principle reason for the widespread use of large eddy simulation models.

3 Cloud physics and chemistry issues

The climatic significance of low layer clouds is strongly dependent on the number, size and constitution of aerosols. Those which are soluble in water and large enough to serve as nuclei for condensing water drops at small supersaturations are called cloud condensation nuclei (CCN). For each nucleus there is a maximum equilibrium supersaturation, which varies inversely with the nucleus size, beyond which the nucleus is activated and grows as a drop of dilute salt solution. The equilibrium supersaturation also varies with the salt of which the aerosol is composed.

Determination of the number of cloud drops which will form in a rising air mass containing a distribution of aerosols of mixed sizes and properties is a somewhat difficult task, with often uncertain results. In general, however, the more nuclei, the more drops are activated. Air masses which have been over the ocean for a considerable time are usually relatively clean, with CCN number <100 cm^{-3}, while those currently or recently over land are usually much dirtier.

The reflectivity of a cloud is largely dependent on its thickness and on the surface area of its drops, so that a cloud formed on a continental aerosol may look considerably brighter from above and darker from below than a marine cloud with the same liquid water content. This has been demonstrated from satellite imagery downstream of urban or rural pollution sources, and more spectacularly from visible ship tracks, as shown in Fig. 2.

The aerosols which appear in the atmosphere have various sources. However, the soluble salts which are most frequently effective as CCN are ammonium sulfate and sodium chloride. Ammonium sulfate is initially generated as very small particles which must coalesce from Brownian motion, probably for several hours, in order to become large enough to act as CCN. Sodium chloride and other ocean salts are introduced by evaporating spray, formed

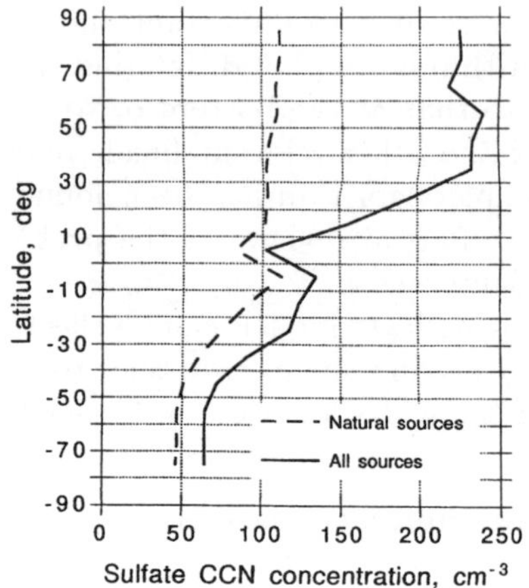

Figure 5. The zonally averaged (oceanic areas only) profiles of the sulfate aerosol concentration. The dashed line shows the contribution from natural sources only, while the solid line shows the contribution from all sources (natural plus anthropogenic). From Z. Kogan et al.[22]

when the surface winds are strong enough to produce whitecaps. Some of the particles produced this way are quite large, greater than 1 μm in radius. They are very effective as CCN and grow to produce rather large drops by condensation.

Figure 5 shows the amount of sulfate in the air currently as well as an estimate of its preindustrial value. The pollutants released from fires and burned hydrocarbons, especially sulfur oxides, tend to produce clouds which are more reflective of sunlight than those formed in cleaner air (Twomey[24], Platnick and Twomey[25]). Albrecht[26] proposed that clouds formed in dirty air are also more durable, because the droplets are less likely to coalesce and rain out. Dirty air without clouds also reflects considerable solar energy. The effect of dry aerosols is usually called the direct effect, while that of clouds forming on them is the indirect effect. Fig. 6 shows estimates made by four different methods of cooling due to the total effect of anthropogenic aerosols. The magnitudes, up to 10 W m^{-2} for methods III and IV are locally larger in parts of the northern hemisphere and, for method IV, globally approach the expected warming due to the greenhouse effect.

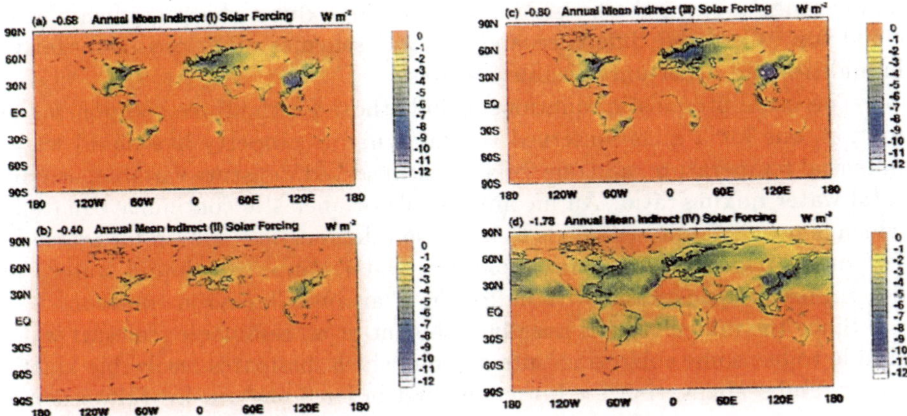

Figure 6. Geographical distribution of the annual mean indirect forcing (W m^{-2}) due (a) method I, (b) method II, (c) method III, (d) method IV. From Kiehl *et al.*[23]

4   LES purposes, methodologies and problems

Large eddy simulation of convectively driven layer clouds was first applied by Deardorff[19], revived by Moeng[27], and actively pursued almost continuously since (e.g., Schumann and Moeng[28], Khairoutdinov and Kogan[29], Stevens *et al.*[30]). The initial purpose was to test mixed layer and other theories, in comparison also with laboratory and observational data. Several comparison tests have been made, in which different LES models were run with essentially the same initial and forcing conditions (Moeng, *et al.*[31], Bretherton *et al.*[32]). Recently a principal goal has been to develop and test single column parameterizations suitable for entry into global climate models (GCMs). Some single-column models were compared by Bretherton *et al.*[33]

LES models of stratiform clouds are restricted to a small domain by most meteorological standards, typically 5–10 km in horizontal extent and a few km in the vertical. The small domain allows for high resolution, usually 50 m horizontal spacing or finer, and 25 m or finer in the vertical, often with the finest spacing near the surface and the mixed layer top. The numerical methodology usually involves either centered 2nd or 4th order accuracy differencing, pseudospectral methods, or more recently one of several higher order schemes with monotonicity properties, that is they conserve the extreme values, e.g., Smolarkiewicz and Grabowski[34]. This is helpful when dealing with moisture variables, which are inherently positive but cover a fairly wide dynamic range even in the lower troposphere and exhibit sharp gradients near cloud top. Centered difference and spectral schemes often predict negative humidities and other unrealizable features.

The usual prognostic equations include those for the three components of motion, one for a quasiconserved thermodynamic property like liquid water potential temperature, and one for a quasiconserved moisture variable, usually total water mixing ratio. An incompressible or anelastic continuity equation, which is usually assumed, leads to a Poisson diagnostic equation for pressure. For models without cloud microphysics, water vapor and liquid water are determined diagnostically with aid of the Clausius–Clapeyron equation.

Since the mixed layer is usually turbulent at all resolvable scales, it is desirable to use some subgrid closure scheme which dissipates resolvable kinetic energy and scalar variances. Either the first order Smagorinsky–Lilly scheme, including a correction for buoyant stability, or the so-called 1.5 order scheme using a turbulent energy equation is most commonly used.

The most serious problems with most LES models are the usual economic ones of having neither enough fine-scale resolution or a large enough domain to simulate important observed structures. The fine-scale resolution limit near cloud top has been found to lead to incorrect predictions of the entrainment rate (Stevens and Bretherton[35]). Most modelers use a horizontal/vertical mesh spacing ratio of 2:1 or a little higher, but enhance only the vertical resolution near the mixed layer boundaries. The domain size is usually sufficient to contain a number of convective elements, but observational data usually show larger scale structures, 20 km or more in extent, often cellular, superimposed and perhaps partially controlling the convective scales.

Several recent model simulations have applied detailed cloud microphysics, including the processes of CCN activation, growth by condensation and coalescence, fallout, evaporation, and drop deactivation. These require consideration of rather wide size ranges of both CCN and drops. Typically distribution functions are created, which would be continuous in theory but are discretized for computation, with bins containing drops within various size ranges. One version, due to Kogan et al.[36], uses 10–20 CCN bins, with sizes chosen with reference to the supersaturation required to activate them, and 20–40 drop bins distributed logarithmically, typically 1 per octave of drop mass. A more sophisticated but computationally burdensome scheme due to Liu[37] uses a two-parameter distribution function, with bins assigned to specific size ranges of each of nucleus mass and drop mass. Since each distribution function variable is advected by the turbulent flow fields as well as participating in various cloud physics processes, the increased computational load is quite heavy for both methods, requiring 30–60 new prognostic variables for

Kogan's method and hundreds for that of Liu.

Cloud chemistry has usually been considered separately, though sometimes with the aid of a 2D cloud resolving or LES model. To my knowledge no LES model for stratified cloud simulation has included both a detailed cloud microphysics and a chemistry module. That is to be desired in the future, to allow realistic restoration of CCN removed by drop coalescence and fallout.

5   Some examples

Here I show a few of the results of two LES models which include important cloud physics and dynamics processes. These were carried out by doctoral students at the University of Oklahoma, under the direction of Prof. Yefim Kogan.

5.1   Khairoutdinov: Removal of CCN by drizzle

The first example is due to Khairoutdinov[38] (abbreviated hence as K), further presented by Khairoutdinov and Kogan[29]. They used the cloud physics scheme due to Kogan *et al.*[36] When the humidity increases to saturation, usually during ascent, a droplet bin is formed and replaces the activated CCN bin. Drops grow into larger bins by condensation and coalescence and shrink back from evaporation. When the drops in a bin evaporate they return as CCN particles. Drop coalescence and drizzle remove droplets and subsequently CCN from the population, as they must in nature. The fault in this model is that the condensed drops carry no memory of the mass of the CCN solute within them. An evaporated drop which has grown by coalescence should leave a larger aerosol particle behind than those from which it was formed, but that information is lost. This loss can be partly restored by enlarging the CCN recovered from evaporating drops so that the total solute mass remains approximately constant, except for what falls to the surface

An experiment carried out by K tested an hypothesis due to Ackerman *et al.*[39], that an active and long-lived cloud layer can remove CCN faster than turbulent diffusion or chemical processes can restore them. As the number of nuclei become smaller, the cloud physics processes become more efficient, so that nearly all cloud drops grow to drizzle sizes and many of them fall out, removing the remaining CCN even faster. Ship-based observers in the eastern Pacific have occasionally found extremely low values of CCN, that is a few per cc, and in such cases they sometimes observed drizzle falling from a nearly clear sky. It is thought that this process may take several days to complete.

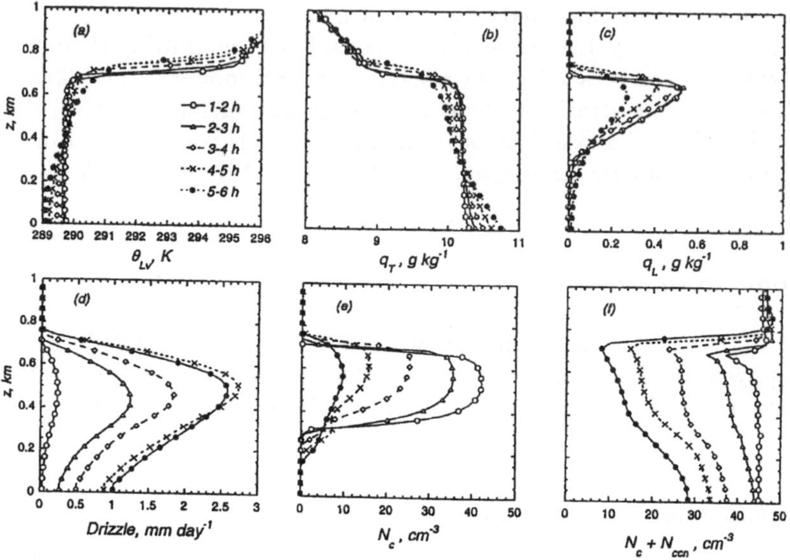

Figure 7. Time evolution of the mean profiles of (a) virtual liquid water potential temperature, (b) total water, (c) liquid water, (d) drizzle rate, (e) drop concentration, and (f) total particle concentration. From Khairoutdinov[38].

The numerical experiment by K simulated this CCN removal process without any addition of new CCN. In the experiment the result of Ackerman *et al.*[39] was mostly obtained in about 6 hours, probably because no new aerosols were created to replace those lost by condensation and fallout. Fig. 7 shows the evolution of several parameters. Looking first at panel (e), the concentration of water drops, one sees that concentration decreases rapidly in the cloud layer, which starts at about 300 m, but increases immediately below it, apparently due to drizzle drops falling through it, most of which evaporate before they hit the surface. Panel (f) shows the sum of water drops and nuclei, which also decreases steadily and rapidly. Initially this quantity is assumed uniform below and within the cloud layer. Condensation and evaporation produce an exchange between drop and nucleus number, with the sum remaining constant, but coalescence and fallout reduce the number. Panel (d) shows the rate of drizzle fall, which increases rapidly for the first four hours. Panel (c), the liquid water mixing ratio, drops fairly rapidly also, but not as fast as the drop number, indicating that the average drop size increases. Panels (a)

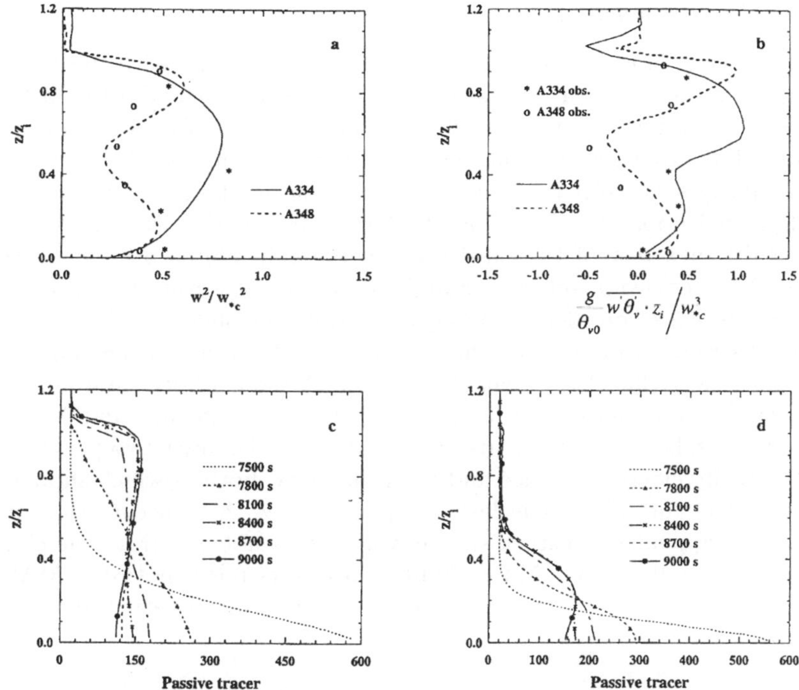

Figure 8. Vertical profiles of the dimensionless vertical velocity variance (a) and buoyancy flux (b) in the A334 case (solid line) and A348 case (dashed line). The stars show the observational data in the A334 case and the circles those for the A348 case. The evolution of the tracer vertical profile in cases A334 and A348 in shown in (c) and (d) respectively. From Liu *et al.*[40]

and (b) are the profiles of virtual liquid water potential temperature, $\theta_{\ell v}$, and total water mixing ratio, q_t. The total water content increases in the subcloud and lower part of the cloud layer and decreases in the upper part, while $\theta_{\ell v}$ does the reverse. Apparently this is due to loss of water in the region where drizzle is forming and gain where it is evaporating, so that the time changes are proportional to the vertical derivative of the curves in panel (d). All the profiles show a steady increase in the depth of the mixed layer of about 100 m over the 6 hour time period, an entrainment rate of 0.5 cm s^{-1}.

5.2 Liu: MAST Simulations

The second example is due to Liu, *et al.*[40] It was also done with the Kogan cloud physics model and was intended to simulate some results of the MAST experiment, which investigated the effects of introduction of large numbers of CCN from ship exhausts into marine air with preexisting cloud cover. The result was fairly successful in showing the increase in cloud droplet number and sunlight reflection after injection of a large number of CCN. Fig. 8 shows partial results of a test of the sensitivity of the ship plume process on the structure of the environment, as found from aircraft observations. In one of the two experiments considered (A334) a well-mixed cloud-topped boundary layer was observed, while in the other (A348) the cloud layer is partially decoupled from the surface by a weak stable layer. Although both environments were observed in the daytime, A334 is more typical of nighttime and A344 of daytime, when solar radiation heats the cloud layer. The first two panels are plots of vertical velocity variance and buoyancy flux, both of which show the decoupling effect for A348. The last two panels show the concentration of a passive scalar equivalent to stack smoke, which does not penetrate the cloud layer in the decoupled case. Many additional results relevant to the MAST experiment and the two-parameter microphysics parameterization are found in Liu[37].

6 Summary

The structure, evolution and interactions of stratified cloud layers are now regarded as important elements in global and local climate dynamics, and also of importance to operational meteorology, particularly for aircraft terminal operations. Mixed layer models provide a relatively simple theoretical framework, but these are incomplete in various ways. The large eddy simulation approach has been applied with some success, with partial inclusion of realistic cloud physics but not yet aerosol chemistry. Two recent LES models are described, with unique features in dynamic structure and cloud microphysics.

References

1. J. L. Hilliker and J. M. Fritsch, *J. Appl. Meteor.* **38**, 1692–1705 (1999).
2. R. A. Brost, D. H. Lenschow and J. C. Wyngaard, *J. Atmos. Sci.* **39**, 800–817 (1982).
3. M. D. King, S.-C. Tsay and S. Platnick, in *Aerosol Forcing of Climate*, R. J. Charlson and J. Heintzenberg, eds., 227–255 (J. Wiley & sons, New York, 1995).

4. S. G. Warren, C. J. Hahn, J. London, R. M. Chervin and R. L. Jenne, NCAR/TN-273+STR (National Center for Atmospheric Research, Boulder, CO, 1986).

5. S. G. Warren, C. J. Hahn, J. London, R. M. Chervin and R. L. Jenne, NCAR/TN-317+STR (National Center for Atmospheric Research, Boulder, CO, 1988).

6. D. A. Randall, B. Albrecht, S. Cox, D. Johnson, P. Minnis, W. Rossow and D. O. C. Starr, *Adv. Geophys.* **38**, 37–177 (1996).

7. B. A. Albrecht, D. A. Randall and S. Nicholls, *Bull. Amer. Meteor. Soc.* **69**, 618–626 (1988).

8. C. S. Bretherton and R. Pincus, *J. Atmos. Sci.* **52**, 2707–2723 (1995).

9. C. S. Bretherton, P. Austin and S. T. Siems, *J. Atmos. Sci.* **52**, 2724–2735 (1995).

10. P. G. Duynkerke, H. Zhang and P. J. Jonker, *J. Atmos. Sci.* **52**, 2763–2777 (1995).

11. D. G. James, *Q. J. Roy. Meteor. Soc.* **85** (1959).

12. A. Slingo, S. Nicholls and J. Schmetz, *Q. J. Roy. Meteor. Soc.* **108**, 833–856 (1982).

13. S. Nicholls and J. Leighton, *Q. J. Roy. Meteor. Soc.* **112**, 431–460 (1986).

14. S. Nicholls and J. D. Turton, *Q. J. Roy. Meteor. Soc.* **112**, 461–480 (1986).

15. P. A. Durkee, K. J. Noone and R. T. Bluth, *J. Atmos. Sci.* **57**, 2523–2541 (2000).

16. D. P. Rogers and J. W. Telford, *Q. J. Roy. Meteor. Soc.* **112**, 481–500 (1986).

17. D. K. Lilly, *Q. J. Roy. Meteor. Soc.* **94**, 292–309 (1968).

18. J. W. Deardorff, *Quart. J. Roy. Meteor. Soc.* **102**, 563–582 (1976).

19. J. W. Deardorff, *J. Atmos. Sci.* **36**, 424–436 (1979).

20. C. S. Bretherton and M. C. Wyant, *J. Atmos. Sci.* **54**, 148–166 (1997).

21. M. C. VanZanten, P. G. Duynkerke and J. W. M. Cuijpers, *J. Atmos. Sci.* **56**, 813–828 (1999).

22. Z. N. Kogan, Y. L. Kogan and D. K. Lilly, *J. Geoph. Res.* **102**, 25927–25939 (1997).

23. J. T. Kiehl, T. L. Schneider, P. J. Krasch and M. C. Barth, *J. Geoph. Res.* **105**, 1441–1457 (2000).

24. S. Twomey, *Atmos. Environ.* **8**, 1251–1256 (1974).

25. S. Platnick and S. Twomey, *J. Appl. Meteor.* **33**, 334–347 (1994).

26. B. A. Albrecht, *Science* **245**, 1227–1230 (1989).

27. C.-H. Moeng, *J. Atmos. Sci.* **43**, 2886–2900 (1986).

28. U. Schumann and C.-H. Moeng, *J. Atmos. Sci.* **48**, 1746–1757 (1991).

29. M. F. Khairoutdinov and Y. L. Kogan, *J. Atmos. Sci.* **56**, 2115–2131 (1999).

30. B. Stevens, C.-H. Moeng and P. P. Sullivan, *J. Atmos. Sci.* **56**, 3963–3984 (1999).

31. C.-H. Moeng, W. R. Cotton, C. Bretherton, A. Chlond, H. G. Peterson, M. Khairoutdinov, S. Krueger, W. S. Lewellen and M. MacVean, *Bull. Amer.*

Meteor. Soc. **77**, 261–278 (1996).

32. C. S. Bretherton, M. K. MacVean, P. Bechtold, A. Chlond, W. R. Cotton, J. Cuxart, H. Cuijpers, M. Khairoutdinov, B. Kosovic, D. Lewellen, C.-H. Moeng, P. Siebesma, B. Stevens, D. E. Stevens, I. Sykes and M. C. Wyant, *Q. J. Roy. Meteor. Soc.* **125**, 391–422 (1999).

33. C. S. Bretherton, S. K. Krueger, M. C. Wyant, P. Bechtold, E. Van Meijgaard, B. Stevens and J. Teixeira, *Bound. Lay. Meteor.* **93**, 341–380 (1999).

34. P. K. Smolarkiewicz and W. W. Grabowski, *J. Comput. Phys.* **86**, 355–375 (1990).

35. D. E. Stevens and C. S. Bretherton, *Q. J. Roy. Meteor. Soc.* **125**, 425–439 (1999).

36. Y. L. Kogan, M. P. Khairoutdinov, D. K. Lilly, Z. N. Kogan and Q. Liu, *J. Atmos. Sci.* **52**, 2923–2940 (1995).

37. Q. Liu, Ph.D. dissertation, University of Oklahoma, 135 pp. (1997).

38. M. Khairoutdinov, Ph.D. dissertation, University of Oklahoma Graduate College, 169 pp. (1997).

39. A. S. Ackerman, O. B. Toon and P. V. Hobbs, *J. Atmos. Sci.* **52**, 1204–1236, (1995).

40. Q. Liu, Y. L. Kogan, D. K. Lilly, D. W. Johnson, G. E. Innis, P. A. Durkee and K. E. Nielsen, *J. Atmos. Sci.* **57**, 2779–2791 (2000).

ON THE MODELING OF DEEP CONVECTIVE CLOUDS OVER MEXICO CITY

G.B. RAGA AND J. MORFIN

Centro de Ciencias de la Atmósfera, UNAM, Ciudad Universitaria, 04510 Mexico DF, Mexico; E-mail:raga@servidor.unam.mx

We have implemented in Mexico the Advanced Regional Prediction System (ARPS) developed by the Center for Analysis and Prediction of Storms (CAPS) of the University of Oklahoma, to study deep convective clouds that develop almost daily over Mexico City during the rainy season (from June to November). Mexico City is located in an elevated basin surrounded by mountains on three sides (East, South and West), that reach up to 1.5 km above mean basin level. The orographic forcing is crucial to the development of convection, but the larger scale conditions determine the location and strength of the convective activity. This convection is more frequently observed in the SW corner of the basin, where climatological precipitation values are a factor of 2 larger than in the NE sector. In this study we have simulated a storm that occurred on August 18, 1997, and gave way to generalized flooding in the WSW area of the basin.

The results indicate that the main storm characteristics are fairly well reproduced by the model, when comparing with the precipitation measured at the surface by the city's raingauge network. The dynamics of a long-lasting precipitating cloud appears to be the interaction of a daughter cell with its parent cells. The main mechanism for precipitation production is the interaction of cloud ice with hail in the initial 2 cells, although some warm rain conversion is also active.

1 Background

Mexico City is located at 19°N and in an elevated basin (at 2.2km above sea level), almost entirely surrounded by mountains. The climate in the Mexico City basin is characterized by a dry season from November to April followed by a rainy season from May to October. Much of the rain during the wet season is associated with convective thunderstorms embedded in the deep moist trade winds.

The average annual precipitation pattern observed within the basin[1] shows a strong gradient from the semiarid regions to the northeast (with 400 mm per year) to the humid southwest corner (that receives 1000 mm per year). This spatial pattern is clearly linked to the orographic influence, which forces convection in the southwest corner of the basin.

The purpose of this study is specifically to test the hypothesis that the orography plays a major role in the development of deep convection and precipitation during the wet season.

122

2 Methodology

As a modeling tool, we utilize the Advanced Regional Prediction System (ARPS) numerical code developed at the Center for the Analysis and Prediction of Storms of the University of Oklahoma[2].

The model is a nonhydrostatic prediction model and is appropriate for use on scales ranging from a few meters to hundreds of kilometers. It is based on the compressible Navier–Stokes equations describing the atmospheric flow. The terrain-following coordinates allow the simulations of flow over complex topography, such as in this case. A variety of physical processes are taken into account in this model, with parameterizations for surface processes, radiation, turbulence and cloud microphysics.

The governing equations of the atmospheric component of the model include momentum, heat (potential temperature), mass (pressure), water substances, turbulent kinetic energy and the equation of state. These equations are represented in a curvilinear coordinate system which is orthogonal in the horizontal. The governing equations used are the result of direct transformation from the Cartesian system and are expressed in a fully conservative form. These equations are solved in a rectangular computational space using finite differences. The computational grid can be arbitrarily defined and the transformation Jacobians are calculated after the numerical grid is defined. The model variables are staggered on an Arakawa C-grid, with scalars defined at the center of the grid boxes and the normal velocity components defined on the corresponding box faces.

Since the model atmosphere described by the governing equations is compressible, meteorologically unimportant acoustic waves are also supported by the model. The presence of acoustic waves severely limits the time step size of explicit time integration schemes. To improve model efficiency, the mode-splitting time integration technique is employed. This technique divides a big integration time step into a number of computationally inexpensive small time steps and updates the acoustically active terms every small time step while computing all the other terms only once every big time step. The large timestep integration uses a centered three-level (leapfrog) time differencing scheme. With the exception of the advection terms, the spatial difference terms are second-order accurate. The advection can be either second- or fourth-order accurate. The model assumes that the top boundary is flat and at a height defined by the user, and that the vertical velocity normal to the boundary vanishes there. There is also a Rayleigh damping term added to the conservation equations of momentum, potential temperature and water quantities near the top of the domain to absorb upward propagating wave disturbances and to eliminate wave reflection at the top boundary. These terms act to damp the perturbations from the base state and the height of this damping layer is in general 1/3 of the total domain depth. The bottom boundary is at a height $z = h(x,y)$, where h is the terrain height. When the terrain is not flat, vertical velocity will not be zero at the ground. The terrain-following coordinate transformation used by ARPS ensures that the computational grid line at the lower boundary follows the terrain. Five types of lateral boundary conditions are available to the user: rigid wall, periodic, zero normal gradient, wave radiating open boundary and externally specified boundary conditions. For the present study, zero normal gradient conditions were used in all lateral boundaries.

Wind components and the state are defined as the sums of base-state variables and the deviations from the base state. The base state is assumed to be horizontally homogeneous, time invariant and hydrostatically balanced. The base state can be initialized using prescribed analytical functions or an external sounding.

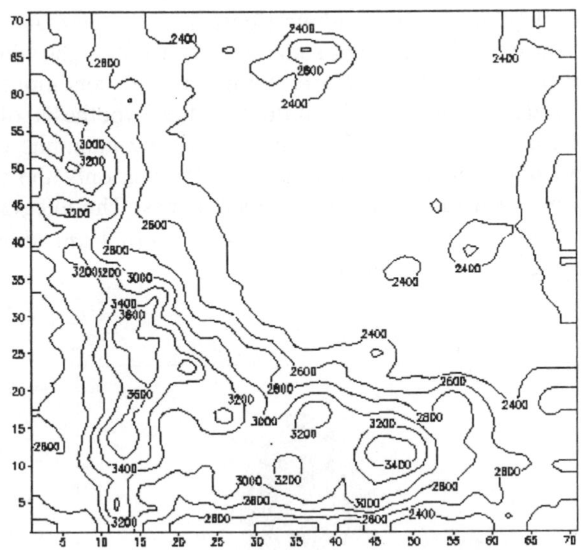

Figure 1. Modeling domain and contours of topography.

Since the model will be used to study cloud development it is relevant to discuss some aspects of the cloud physics parameterizations. The Kessler warm rain microphysics parameterization considers three categories of water: water vapor, cloud water and rain water. Each of the liquid forms is implicitly characterized by a droplet distribution. Small cloud droplets are first formed when the air becomes saturated and condensation occurs. If the cloud water mixing ratio exceeds a threshold value, raindrops are formed by autoconversion from the cloud droplets. The raindrops then collect smaller cloud droplets by accretion as they fall at their terminal speed. If cloud droplets enter unsaturated air they evaporate until either the air is saturated or until the droplets are exhausted. Raindrops also evaporate in a subsaturated environment at a rate depending on their concentration and the saturation deficit. When the ice phase is present, many more processes will be involved. Negative water quantities produced by advection and by the vertical flux terms associated with rainwater fallout are not adjusted. The negative values are set to zero and only the positive values are used in the microphysical calculations. Since both positive and negative values are involved in the advection and mixing processes, the total water content is conserved apart from the rainwater fallout.

The environmental conditions used to initialize the model correspond to the 6 am sounding (local time) on August 18, 1997. Very deep convection was observed to develop this day and intense precipitation over the slopes caused washouts and street flooding in the southwest corner of the city. A deep layer of instability was evident from the morning sounding, with very light winds (both the zonal and meridional components). These environmental conditions (with little evidence of large scale forcing) are ideal to test the hypothesis of orographic forcing. Convection is *not* forced in these simulations by including a hot bubble (not even by 'white noise' in the temperature field); it is only the result of the differential heating between sun-facing and shaded slopes. The integration domain covers 68 km × 68 km × 18 km, with a spatial resolution of 1 km, shown in Fig. 1 together with the topography in the region.

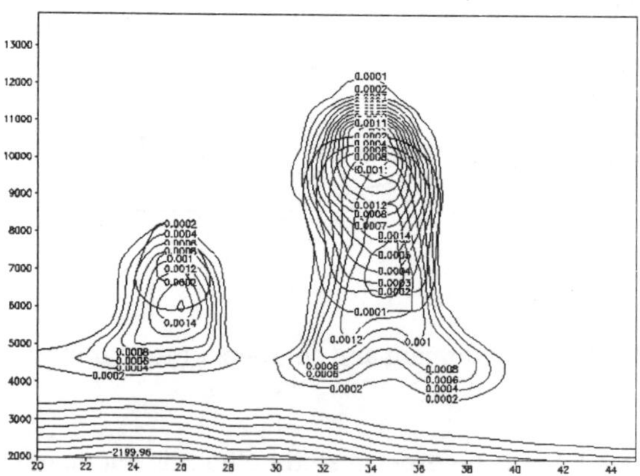

Figure 2. Vertical cross section of cloud water and cloud ice mixing
ratios at 11:20 am (local time)

3   Results

Figure 2 shows the vertical cross section of the simulated cloud water and ice water fields after more than 5 hours of integration, corresponding to 11:20 am local time. This cross section corresponds to a vertical plane through the ridge shown in the SW corner in Figure 1. Two convective cells are observed, at different times during their evolution.

Ten minutes later (Fig. 3) these two clouds ('parent cells') appear to merge at cloudbase, while continuing their vertical development. The new cloud ('daughter cell') is formed at this point in time of merely cloud water. The clear gap between the parent cells above 7 km is due to the downward motions at the edges of the rising thermals. This downdraft is preventing the development of the incipient cloud between the larger cells. Figure 4 presents the streamlines along the plane shown in Fig. 3, where this can be easily observed.

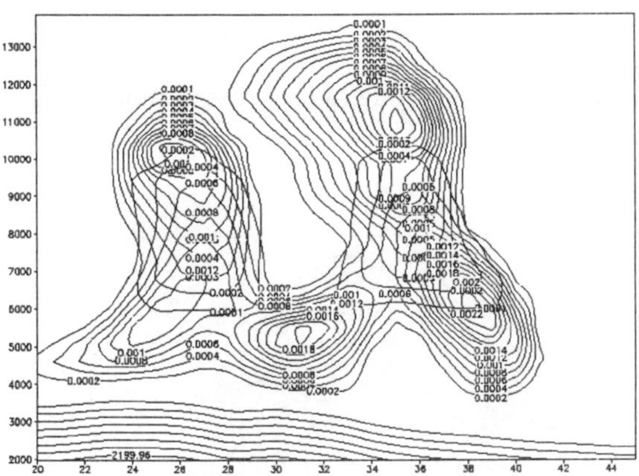

Figure 3. Same as Fig. 2 but at 11:30 am (local time)

Only when the larger cells start to decay after fallout of heavy precipitation, can the daughter cell begin to develop. This daughter cell appears to grow and produce precipitation somewhat faster than the parent cells and we hypothesize that the interaction with the parent cells is inducing the faster development. Hail and rain fall out of the parent cells into the growing daughter cell and accelerate the production of hydrometeors in the latter. This interaction results in as much rain water and hail in the daughter cell as in the parent cells after only 20 minutes of development, compared to more than 30 minutes in the parent cells, as can be observed in Fig. 5.

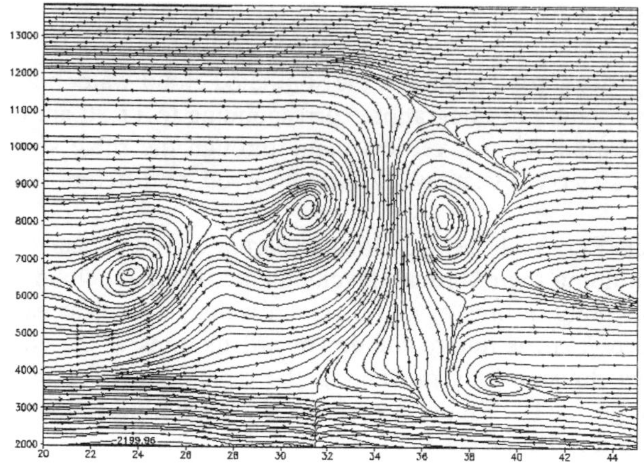

Figure 4. Streamlines corresponding to the cross section in Fig. 2.

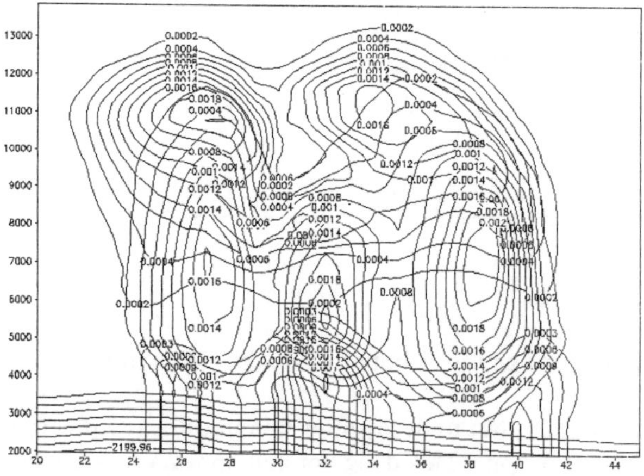

Figure 5. Same as Fig. 2 but for snow, hail and rain water mixing ratios.

To gain insight into whether the precipitation was produced mainly by warm rain or by cold rain processes, we have isolated the hydrometeor interactions that lead to rain water by either process, and are presented in Figs. 6 and 7 for the 20 minute period under discussion. In the daughter cell, the acumulated precipitation is dominated by the warm rain process, due to the fact that falling rain from the parent cells is accelerating its production.

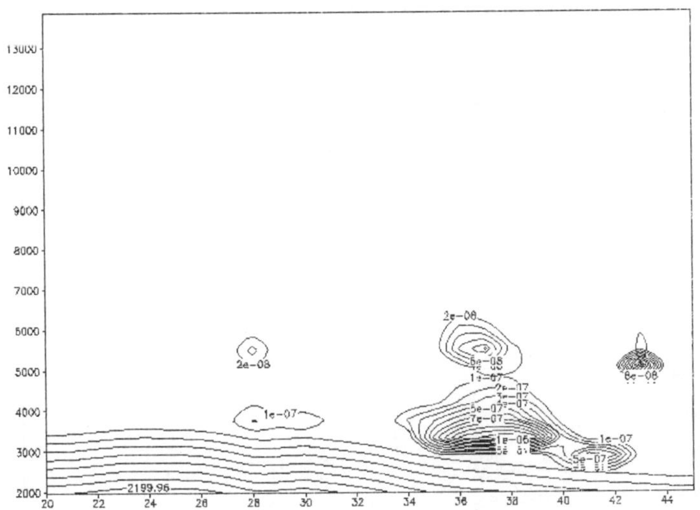

Figure 6. Production of rain by warm rain processes in a 20 minute period.

4 Conclusions

These preliminary results indicate that indeed the topographic forcing alone can be responsible for the convective activity observed in the southwest corner of the Mexico City basin. The main precipitation mechanism appears to be cloud ice–hail interaction present in the initial 2 cells, although some warm rain conversion is also active. This interaction between cells seems to be responsible for the prolonged period of precipitation, causing a large accumulation. Qualitatively, this result is in agreement with the observed pattern of precipitation at the surface on Aug.18, 1997.

128

5 Acknowledgements

The simulations were made using the Advanced Regional Prediction System (ARPS) developed by the Center for Analysis and Prediction of Storms (CAPS), University of Oklahoma. CAPS is supported by the National Science Foundation and the Federal Aviation Administration under grant ATM92-20009. Partial support was obtained from grant SC005996 of the UNAM–SGI R&D Grant Program.

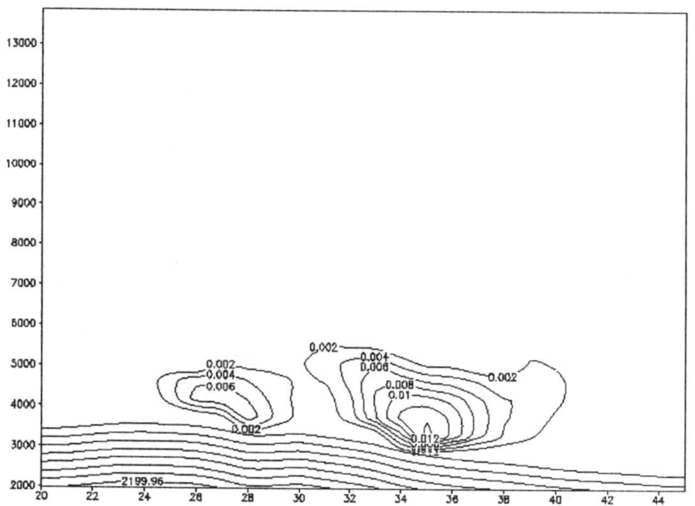

Figure 7. Same as Fig. 6 but by cold rain processes.

References

1. E. Jauregui and E. Romales, "Urban effects on convective precipitation in Mexico City", *Atmos. Environ.* **30**, 3383–3389 (1996).
2. M. Xue, K. Droegemeier and V. Wong, "The advanced regional prediction system and real-time storm weather prediction", *Proc. of the Internat. Workshop on limited-area and variable resolution models* (WMO, Beijing, 1995).

SPECTRAL STRUCTURE OF GROWING NORMAL MODES FOR EXACT SOLUTIONS TO THE BAROTROPIC VORTICITY EQUATION ON A SPHERE

YURI N. SKIBA

Centro de Ciencias de la Atmósfera, UNAM, 04510, México, D.F., México
E-mail: skiba@servidor.unam.mx

ANDREI Y. STRELKOV

The Keldysh Institute of Applied Mathematics, RAS, Moscow, A-47, RUSSIA
E-mail: strelkov@caravan.ru

The exponential instability of the Legendre polynomial flows, Rossby–Haurwitz waves, Wu–Verkley waves and monopole, dipole and quadrupole modons on a sphere is considered. These flows are exact solutions to the barotropic vorticity equation on a rotating sphere. We have obtained a conservation law for disturbances and a condition necessary for the normal mode instability of each such flow, and estimated the maximum growth (and decay) rate of the modes. We have also shown that the amplitude of any unstable, decaying and nonstationary mode is orthogonal to the basic solution in the energy inner product. Any mode not satisfying the new instability condition is neutral. The condition localizes the unstable modes in phase space, characterizes the spectral structure of growing perturbations, and is helpful in testing a computational algorithm developed for the numerical linear stability study of a flow on a sphere. Some properties of the conditions for different types of modons are discussed, too.

1 Introduction

The motion of an ideal incompressible fluid on a rotating sphere S is described by the barotropic vorticity equation (BVE)

$$\Delta \psi_t + J(\psi, \Delta \psi + 2\mu) = 0 \tag{1}$$

where $J(a, b) = a_\lambda b_\mu - a_\mu b_\lambda$ is the Jacobian, λ is the longitude, μ is the sine of latitude, and $\Delta \psi$ and $\Delta \psi + 2\mu$ are the relative and absolute vorticity, respectively (hereinafter f_λ, f_μ and f_t denote the partial derivatives of f). Equation (1) also describes quite well the large-scale dynamics of an inviscid and unforced barotropic atmosphere[1−6]. Therefore the stability of exact solutions to (1) is of great hydrodynamic and meteorological interest, and new results in this area can provide insight into deeper understanding of the low-frequency atmosphere variability and climate predictability. In spite of many investigations, the stability problem is still far from its complete resolution. It will suffice to mention that for half a century, the only useful and easily applied linear instability condition for solutions to Eq. (1) has been Rayleigh–Kuo's condition[7]. However, its utility is rather limited. First, it serves only for zonal flows. Second, each Legendre polynomial (LP) flow of degree $n \geq 3$ satisfies this condition if its amplitude is large enough. Third, the Rayleigh–Kuo condition is only the necessary condition for the instability, and its fulfillment does not guarantee the instability[6]. And finally, it provides no information about the growth rate and time–space structure of unstable disturbances. It should also be noted that no instability condition has been developed up to now for Rossby–Haurwitz (RH) waves, Wu–Verkley (WV) waves, or modons. Thus, for a fundamental understanding the instability of flows on a sphere (or at least BVE solutions), an active search for other instability conditions is required. In this work, we consider the linear instability of the following well-known real exact solutions to Eq. (1):

I. A Legendre polynomial (LP) flow

$$\psi(\mu) = a P_n(\mu) \tag{2}$$

where a is the flow amplitude, and $P_n(\mu) = (2^n n!)^{-1} \frac{d^n}{d\mu^n}(\mu^2 - 1)^n$ is the Legendre polynomial of degree n . Mathematically, LP flows are interesting as an orthogonal basis for the zonal flows on a sphere[6].

II. A steady Rossby–Haurwitz (RH) wave

$$\psi(\lambda, \mu) = -\omega\mu + \sum_{m=-n}^{n} \psi_n^m Y_n^m(\lambda, \mu) \tag{3}$$

where $\omega = 2/(\chi_n - 2)$, $\chi_n = n(n+1)$, ψ_n^m is arbitrary amplitude, $Y_n^m(\lambda, \mu) = P_n^m(\mu)\exp(im\lambda)$ is the spherical harmonic of the degree n and zonal number m, and $P_n^m(\mu) = (1 - \mu^2)^{m/2}\frac{d^m}{d\mu^m}P_n(\mu)$ is the associated Legendre function $(n \geq 1)$ [8].

III. A steady (antisymmetric around the equator) Wu–Verkley (WV) wave[1]

$$\psi(\lambda, \mu) = \begin{cases} X_i(\lambda, \mu) - \omega_i\mu + D_i, & \text{in } S_{\text{in}} \\ X_o(\lambda, \mu) - \omega_o\mu + D_o, & \text{in } S_{\text{out}} \end{cases} \tag{4}$$

where $S_{\text{in}} = \{(\lambda, \mu) \in S : \mu \in (-\mu_0, \mu_0)\}$ and $S_{\text{out}} = S \setminus S_{\text{in}}$ are the inner (tropical) and outer (polar) regions, $0 < \mu_0 < 1$, and D_i and D_o are constants. Furthermore, $\omega_i = 2/(\chi_\alpha - 2)$, $\omega_o = 2/(\chi_\sigma - 2)$, α and σ are the degrees of the spherical harmonics representing solution (4) in regions S_{in} and S_{out}, and $X_i(\lambda, \mu)$ and $X_o(\lambda, \mu)$ are the Laplacian eigenfunctions corresponding to the eigenvalues $\chi_\alpha = \alpha(\alpha + 1)$ and $\chi_\sigma = \sigma(\sigma + 1)$, respectively: $-\Delta X_i = \chi_\alpha X_i$, $-\Delta X_o = \chi_\sigma X_o$.

IV. A steady (monopole or dipole) modon by Verkley[2-4] and quadrupole modon by Neven[5]. In the primed system of coordinates (λ', μ'), whose pole coincides with the modon center and is on the latitude circle $\mu = \mu_0$, the modon can also be written in the form (4), but the inner and outer modon regions are defined as $S_{\text{in}} = \{(\lambda', \mu') \in S : \mu' > \mu_a\}$ and $S_{\text{out}} = \{(\lambda', \mu') \in S : \mu' < \mu_a\}$, that is, $\mu' = \mu_a$ is the circle separating S_{in} and S_{out}.[2]

Note that solutions (2)–(4) are of both hydrodynamic and meteorological interest[1-5]. Infinitely smooth LP flows and RH solutions of low degree represent planetary waves on the earth. The WV waves and modons are not so smooth as the LP flows and RH waves, and generally considered as weak BVE solutions. While WV solutions can be used to describe small-scale geometric structures in the low latitudes of the atmosphere, modons are interesting as prototypes of moving isolated vortices and blockings.

2 Some integral formulas involving the Jacobian

We define the inner product and norm in the Hilbert space of square integrable functions on S as

$$\langle \psi, h \rangle = \int_S \psi \overline{h} \, dS \qquad \text{and} \qquad \|\psi\| = \langle \psi, \psi \rangle^{1/2} \tag{5}$$

Note that the boundary between S_{in} and S_{out} is a streamline ($\psi_{\lambda'}(\lambda', \mu_a) = 0$) for the modon, or the union of two streamlines ($\psi_\lambda(\lambda, -\mu_0) = 0$, and $\psi_\lambda(\lambda, \mu_0) = 0$) for the WV wave. The modon inner region S_{in} is a sphere domain bounded by the latitudinal circle $\mu' = \mu_a$. As to the WV wave, its region S_{in} is a latitudinal channel $-\mu_0 \leq \mu \leq \mu_0$ on the sphere S. Taking into account this fact we now give a few integral relations widely used in this work. Let $G = \{(\lambda, \mu) \in S : \mu \in (\mu_a, 1]\}$ be a part of the sphere S bounded by a latitudinal circle $\mu = \mu_a$ $(-1 < \mu_a < 1)$. Then

$$\int_G J(\psi, f) f \, dS = 0 \tag{6}$$

for any complex smooth functions ψ and f such that $\psi_\lambda(\lambda, \mu_a) = 0$. Equation (6) is also valid if $G = \{(\lambda, \mu) \in S : \mu \in [\mu_a, \mu_b]\}$ is a periodic channel on S and the normal velocity component is zero at the walls: $\psi_\lambda(\lambda, \mu_a) = \psi_\lambda(\lambda, \mu_b) = 0$.

If, in addition, ψ is a real function then

$$\Re \int_G J(\psi, f) \overline{f} \, dS = 0 \tag{7}$$

132

where $\Re$ denotes the real part of the number. Also we note that

$$\langle J(f,\mu), f\rangle = 0 \qquad \text{and} \qquad \langle J(f,\mu), \Delta f\rangle = 0 \tag{8}$$

for any complex smooth function f on the sphere S.

3 Conservation laws for disturbances

An infinitesimal perturbation $\psi'(\lambda,\mu,t)$ of the LP flow ψ is governed by the linear equation

$$\Delta\psi'_t + J(\psi, \Delta\psi' + \chi_n\psi') + 2J(\psi',\mu) = 0 \tag{9}$$

For the solution (3) or (4), such perturbations are governed by the equation

$$\Delta\psi'_t + J(\psi, \Delta\psi' + q\psi') = 0 \tag{10}$$

where $q = \chi_n$ for the RH wave, and

$$q(x) = \begin{cases} \chi_\alpha & , \text{ if } x \in S_{\text{in}} \\ \chi_\sigma & , \text{ if } x \in S_{\text{out}} \end{cases} \tag{11}$$

for the WV wave and modon. Let solution ψ be the LP flow (2) or RH wave (3). Taking the inner product (5) of Eq. (9) or (10) with $\Delta\psi' + \chi_n\psi'$ and using formulas (7) and (8) for $G = S$ we get[6]

Proposition 1. *An infinitesimal perturbation to the LP flow (2) or steady RH wave (3) evolves in such a way that its energy and enstrophy decrease, remain constant or increase simultaneously according to*

$$\{\eta(t) - \chi_n K(t)\}_t = \{[\chi(t) - \chi_n]K(t)\}_t = 0 \tag{12}$$

Here $\chi_n = n(n+1)$, and $K(t) = \frac{1}{2}\|\nabla\psi'(t)\|^2$, $\eta(t) = \frac{1}{2}\|\Delta\psi'(t)\|^2$ and $\chi(t) = \eta(t)/K(t)$ are the kinetic energy, enstrophy and square of Fjörtoft's average spectral number[9] of the perturbation ψ', respectively. By (12), while the energy $K(t)$ grows, the number $\chi(t)$ characterizing the geometric scale of the unstable disturbance tends to the number χ_n. Note that (12) is also valid for any real (not necessarily small) perturbation of the LP flow or RH wave[8,10].

Let now solution $\psi(\lambda,\mu)$ be a steady WV wave or modon (4). Taking the inner product (5) of equation (10) with $\Delta\psi' + q\psi'$ and using (11) and (7) we obtain another conservation law[11]:

Proposition 2. *Disturbances to a WV wave (4) or modon are governed by*

$$\left\{\chi_\alpha^{-1}\eta^{(i)} + \chi_\sigma^{-1}\eta^{(o)} - K\right\}_t = 0 \tag{13}$$

where $K(t)$ is the perturbation energy, and

$$\eta^{(i)}(t) = \frac{1}{2}\int_{S_{\mathrm{in}}} |\Delta\psi'|^2\, dS \qquad \text{and} \qquad \eta^{(o)}(t) = \frac{1}{2}\int_{S_{\mathrm{out}}} |\Delta\psi'|^2\, dS \tag{14}$$

are the parts of the perturbation enstrophy concentrated in solution regions S_{in} and S_{out}.

Note that in the particular case when $\chi_\alpha = \chi_\sigma = n(n+1)$ the WV wave is transformed to the RH wave, and (13) is reduced to (12). The laws (12) and (13) are analogous to those obtained earlier for infinitesimal perturbations to a wavy basic state on the β plane[12] and β plane modon[13,14]. They represent a constraint on the evolution of the perturbation energy and enstrophy, and are nothing other than the conservation of pseudoenergy[15]. We will use these laws for deriving necessary instability conditions in section 5.

4 Invariant sets of disturbances

Defining by

$$\delta(t) = \eta^{(o)}(t)/\eta(t) \qquad \text{and} \qquad 1 - \delta(t) = \eta^{(i)}(t)/\eta(t) \tag{15}$$

the fractions of the perturbation enstrophy concentrated in S_{out} and S_{in} ($0 \leq \delta \leq 1$), we can write both (12) and (13) as a conservation law

$$\left\{[p(\psi') - 1]\,K(t)\right\}_t = 0 \tag{16}$$

where $p(\psi') \equiv p(t)$ is a spectral number characterizing the scale of the disturbance $\psi'(\lambda, \mu, t)$ on S; in addition,

$$p(\psi') = \chi(\psi')\chi_n^{-1} = \chi(\psi')\,[n(n+1)]^{-1} \tag{17}$$

for the LP flow or RH wave, and

$$p(\psi') = \chi(\psi')\left\{\delta\chi_\sigma^{-1} + (1-\delta)\chi_\alpha^{-1}\right\} \tag{18}$$

for the WV wave or modon. Due to (16), $p(\psi') \to 1$ when $K(t)$ grows, and all infinitesimal perturbations $\psi'(\lambda, \mu, t)$ to either solution are divided into three invariant sets:

$$M_+ = \left\{\psi' : p(\psi') > 1\right\}, \; M_0 = \left\{\psi' : p(\psi') = 1\right\}, \; M_- = \left\{\psi' : p(\psi') < 1\right\} \tag{19}$$

Note that all the real (not necessary small) perturbations of the LP flow or RH wave are divided into sets (19)[16]. Set M_0 separates the large-scale perturbations of M_- and small-scale perturbations of M_+.

134

5 Exponential instability conditions

Infinitesimal perturbations to solution ψ of Eq. (1) will be searched as a normal mode

$$\psi'(\lambda, \mu, t) = \Psi(\mu) \exp\{im\lambda + \nu t\} \tag{20}$$

if ψ is the LP flow or monopole (zonal) modon, and as a normal mode

$$\psi'(\lambda, \mu, t) = \Psi(\lambda, \mu) \exp\{\nu t\} \tag{21}$$

if ψ is the RH wave, WV wave, and dipole or quadrupole modon. Here m is the zonal wavenumber, and $\nu = \nu_r + i\nu_i$ is complex. The time dependence of the mode is so simple that the spectral number $\chi(\psi') \equiv \eta(t)/K(t)$ of the mode is time-independent and coincides with the spectral number χ_Ψ of the mode amplitude Ψ :

$$\chi(\psi') = \chi_\Psi = \eta_\Psi/K_\Psi \tag{22}$$

Here η_Ψ and K_Ψ are the enstrophy and energy of the mode amplitude. Therefore, $p(\psi')$ is also time-independent, and (16) can be written as

$$\nu_r \left[p(\psi') - 1 \right] K_\Psi = 0 \tag{23}$$

Since $\nu_r \neq 0$ for the growing or decaying modes, we obtain that

$$p(\psi') = 1 \tag{24}$$

is the necessary condition for the exponential instability. Obviously, the amplitude of a decaying mode must also satisfy condition (24). Thus we proved two assertions:

Proposition 3. *Let $n > 2$, $|m|\, o < n$ and $m \neq 0$. A mode (20) of the LP flow (2) is unstable (or decaying) only if spectral number $\chi_\Psi = \eta_\Psi/K_\Psi$ of the mode amplitude $\Psi(\mu)$ is equal to $\chi_n = n(n+1)$.*

Proposition 4. *Let ψ be a steady RH wave (3), WV wave (4), or modon. A mode of the flow ψ is unstable (or decaying) only if*

$$\chi_\Psi = \begin{cases} n(n+1) & , \quad \text{if } \psi \text{ is the RH wave} \\ \left\{ \delta\chi_\sigma^{-1} + (1-\delta)\chi_\alpha^{-1} \right\}^{-1} & , \text{ if } \psi \text{ is the WV wave or modon} \end{cases} \tag{25}$$

where $\chi_\Psi = \eta_\Psi/K_\Psi$ is the spectral number of the mode amplitude $\Psi(\lambda, \mu)$.

Note that any mode (20) of the LP flow (2) is stable if $|m| > n$ or $m = 0$.[17] Due to Propositions 3 and 4, no unstable mode of the LP flow or RH wave can be disclosed by means of computations while the triangular truncation number for the disturbance series of the spherical harmonics is taken less than or equal to n.[6] For the LP flow or RH wave, the instability equation $\chi_\Psi = n(n+1)$ depends only on the basic flow degree n, and the locus of its solutions is a hypersurface (a set of measure zero) in the perturbation phase space (of Fourier coefficients). Any mode whose amplitude does not satisfy this equation is neutral. Unlike this, the necessary condition (25) for the WV wave (4) (or modon) depends not only on the basic flow degrees χ_α and χ_σ, but also on the perturbation energy distribution in the regions S_{in} and S_{out} (i.e., on parameter δ). As a result, the set of unstable disturbances in this case is much more complicated representing a one-parameter family (of δ). In all cases the set of unstable modes is something special, and hence good precision is required of any numerical algorithm used to compute the modes. In this connection, the instability conditions are useful for disclosing various errors in the computer programs developed for the exponential stability study of arbitrary flow on a sphere. These conditions also characterize the spatial (geometric) structure of a growing perturbation for the flows under consideration. Note that Eq. (23) is the requirement that the pseudoenergy associated with growing or decaying normal modes vanishes, as is required for its simultaneous conservation and exponential growth[18].

6 Peculiarities of the instability conditions

We now consider some properties of the instability condition (25) for different BVE solutions[11].

Example 1. *Nonlocal BVE solutions.* Let ψ be a WV wave, or a modon by Verkley[3] or Neven[5], and hence $\chi_\alpha > 0$ and $\chi_\sigma > 0$. Due to (25), χ_Ψ^{-1} is the linear interpolation of χ_σ^{-1} and χ_α^{-1}, and spectral number χ_Ψ of the amplitude of a growing (or decaying) mode is always between χ_σ and χ_α. In particular, if enstrophy η_Ψ of the amplitude of an unstable mode is concentrated only in the inner solution region S_{in} (or only in its outer region S_{out}) then $\delta = 0$ ($\delta = 1$) and $\chi_\Psi = \chi_\alpha$ ($\chi_\Psi = \chi_\sigma$).

Example 2. *Modons with uniform absolute vorticity.* Let ψ be a modon with uniform absolute vorticity in the inner region S_{in}.[4] According to Verkley's theorem ([4], Appendix B), vorticity $\Delta\Psi$ of the amplitude of each unstable (or decaying) mode is zero in S_{in} ($\delta = 1$), and condition (25) is reduced to $\chi_\Psi = \chi_\sigma$. Thus, the normal mode instability condition for such a modon depends only on the modon degree σ in outer region S_{out}, and in this sense resembles those for the LP flow and RH wave.

Example 3. *Isolated modons.* Let ψ be a localized modon[2] of complex degree $\sigma = -0.5 + ik$, where $\chi_\sigma = \sigma(\sigma + 1) = -(k^2 + 1/4) < 0$ and $\chi_\alpha = \alpha(\alpha + 1) > 0$. By Proposition 4, mode (21) of the modon can be unstable only if $\chi_\Psi^{-1} = (1 - \delta)\chi_\alpha^{-1} - \delta |\chi_\sigma|^{-1}$. Since $\chi_\Psi^{-1} > 0$, there is a restriction from above on the fraction δ of the perturbation enstrophy concentrated in region S_{out}: $0 \le \delta < \delta_{cr} = |\chi_\sigma| (\chi_\alpha + |\chi_\sigma|)^{-1} < 1$. Thus, δ_{cr} decreases as the modon degree α grows. In particular, if the amplitude vorticity $\Delta\Psi$ of a mode is equal to zero in the inner modon region S_{in} ($\delta = 1$) then the mode is neutral. Unlike this, any mode to the Example 2 modon with $\delta = 1$ satisfies the necessary instability condition $\chi_\Psi = \chi_\sigma$. Further, due to (25), $\chi_\Psi = \chi_\alpha (1 - \delta/\delta_{cr})^{-1} \ge \chi_\alpha$ for unstable modes of an isolated modon, and the minimum $\chi_\Psi = \chi_\alpha$ corresponds to the case $\delta = 0$ when enstrophy $\Delta\Psi$ of the mode amplitude is zero in the outer modon region.

7   Bounds on the growth rate of the modes

The following two statements estimate the maximum possible growth (decay) rate of unstable modes of the solutions under consideration[6,11]:

Proposition 5. *The maximum growth (decay) rate of a normal mode of the LP flow (2) or stationary RH wave (3) is bounded by*

$$|\nu_r| \le \sqrt{n(n + 1)} \max_S |\nabla\psi| \tag{26}$$

Proposition 6. *The maximum growth (decay) rate of a normal mode of the stationary WV wave (4) or modon is bounded by*

$$|\nu_r| \le \max\{\chi_\alpha, |\chi_\sigma|\} \ \{\delta\chi_\sigma^{-1} + (1 - \delta)\chi_\alpha^{-1}\}^{1/2} \max_S |\nabla\psi| \tag{27}$$

Thus, the mode growth (decay) rate decreases with velocity $\mathbf{u} = \mathbf{k} \times \nabla\psi$ and degree n (or α and σ) of the solution. For instance, from two LP flows (or RH waves) with the same velocity maximum, the flow with smaller n will appear more stable. Estimate (27) leads to (26) under $\chi_\alpha = \chi_\sigma = \chi_n$.

8   Orthogonality of a mode to basic flow

Taking the inner product (5) of Eq. (9) or (10) with the basic flow and using formulas (7), (20) and (21) we obtain

$$\nu \langle \Delta\Psi, \psi \rangle = 0 \tag{28}$$

where Ψ is the mode amplitude, ψ is the basic flow, and $\nu = \nu_r + i\nu_i$. Thus,

$$\langle \Psi, \psi \rangle_K \equiv \langle \nabla \Psi, \nabla \psi \rangle = - \langle \Delta \Psi, \psi \rangle = 0 \tag{29}$$

for any nonneutral ($\nu_r \neq 0$) and nonstationary ($\nu_i \neq 0$) mode. We recapitulate this result in

Proposition 7. *Let ψ be the LP flow, RH wave, WV wave, or modon. Then amplitude Ψ of each unstable, decaying, or nonstationary mode is orthogonal to the basic flow ψ in the energy inner product $\langle \cdot, \cdot \rangle_K$.*

Equation (29) means that integrally (over the whole sphere S), the basic flow velocity $\mathbf{u} = \mathbf{k} \times \nabla \psi$ is orthogonal to the mode amplitude velocity $\mathbf{U} = \mathbf{k} \times \nabla \Psi$:

$$\langle \mathbf{U}, \mathbf{u} \rangle = \int_S \mathbf{U} \cdot \mathbf{u} dS = \int_S (\nabla \Psi \cdot \nabla \psi)\, dS = - \langle \Delta \Psi, \psi \rangle = 0 \tag{30}$$

9 Results

Conservation laws for infinitesimal perturbations to the LP flow, RH wave, WV wave, and modons by Verkley and Neven are derived (Propositions 1 and 2) and used to obtain necessary conditions for their exponential (normal mode) instability. These conditions impose a strict restriction on the spectral energy distribution of unstable (and decaying) modes (Propositions 3 and 4). Namely, Fjörtoft's average spectral number of the amplitude of each unstable mode must be equal to a special number. This number depends just on the basic flow degree (LP flow and RH wave), or on the flow degrees in S_{in} and S_{out} and perturbation energy distribution in these regions. We have estimated bounds of the growth (or decay) rate of the normal modes, too (Propositions 5 and 6). It can also be shown that the amplitude of each unstable, decaying, or nonstationary mode is orthogonal to the solution in the energy inner product (Proposition 7)[6,11]. With the exception of a finite or infinite sums of the Legendre polynomials, our method allows to analyze the stability of four basic sets of exact BVE solutions known up to now. It can also be applied to the stability of asymmetric WV waves[19] and modons by Tribbia[20]. In the latter case, the additional term ψ_t used by Tribbia in (1) must be taken into account.

The series truncation errors accompanying the application of the spectral method in the numerical stability study rise the spectral approximation problem[21]. The new instability conditions provide exact value of the spectral number χ_Ψ for each unstable mode, and thus allow testing the computational numerical stability study algorithms (disclosing various errors in the computer programs)[22–24]. In particular, the corresponding condition for a modon helps to control the quality of calculations. Indeed, the derivatives of the modon vorticity are not continuous on a sphere at the boundary between S_{in} and S_{out}, and as a result, Fourier series for the modon and its perturbation converge slowly near by this boundary (the Gibbs phenomenon). As a result, condition (25) for an unstable mode of the modon is fulfilled just approximately, besides, the difference between the theoretical and numerical values of χ_Ψ decreases slowly as the resolution (series truncation number) increases[22]. Thus, a high resolution is required to confide in the numerical modon stability results[23,24]. This note is true for the WV wave, as well. Also note that for the LP flow, the new instability condition complements the well-known Rayleigh–Kuo condition[7] in the sense that while the latter relates to the basic flow structure, the former refers to the perturbation structure.

Acknowledgments

This work was supported by PAPIIT grant IN122098 (UNAM), CONA-CyT project 32247-T (Mexico), Silicon Graphics–UNAM research project SC-011800, and SNI grant (Mexican National System of Investigators).

References

1. P. Wu and W. T. M. Verkley, *Geophys. Astrophys. Fluid Dynamics* **69**, 77 (1993).
2. W. T. M. Verkley, *J. Atmos. Sci.* **41**, 2492 (1984).
3. W. T. M. Verkley, *J. Atmos. Sci.* **44**, 2383 (1987).
4. W. T. M. Verkley, *J. Atmos. Sci.* **47**, 727 (1990).
5. E. C. Neven, *Geophys. Astrophys. Fluid Dynamics* **65**, 105 (1992).
6. Yu. N. Skiba, *Geophys. Astrophys. Fluid Dynamics*, **92**, 115 (2000).
7. H.-L. Kuo, *J. Meteorology* **6**, 105 (1949).
8. Yu. N. Skiba, *Sov. J. Numer. Analys. Math. Modelling* **6**, 515 (1991).
9. R. Fjörtoft, *Tellus* **5**, 225 (1953).
10. M. Subbiah and M. Padmini, *Indian J. Pure Appl. Math.* **30**, 1261 (1999).
11. Yu. N. Skiba and A. Y. Strelkov, *Geophys. Astrophys. Fluid Dynamics* **93**, 39 (2000).
12. D. G. Andrews, *J. Atmos. Sci.* **40**, 85 (1983).
13. E. W. Laedke and K. H. Spatschek, *Phys. Fluids* **29**, 133 (1986).
14. G. E. Swaters, *Phys. Fluids* **29**, 1419 (1986).
15. T. G. Shepherd, *Advances in Geophysics* **32**, 287 (1990).
16. Yu. N. Skiba, *Atmósfera* **6**, 87 (1993).
17. Yu. N. Skiba and J. Adem, *Numer. Meth. Part. Differ. Equations* **14**, 649 (1998).

18. P. Ripa, *Pure and Applied Geophysics* **133**, 713 (1990).

19. P. Wu, *On Nonlinear Structures and Persistent Anomalies in the Atmosphere*, Ph.D. Thesis, 168 pp. (Imperial Colledge, London, 1992).

20. J. J. Tribbia, *Geophys. Astrophys. Fluid Dynamics*, **30**, 131 (1984).

21. Yu. N. Skiba, *Numer. Meth. Part. Differ. Equations* **14**, 143 (1998).

22. I. Pérez-Garcia and Yu. N. Skiba, Tests of a numerical algorithm for the linear instability study of flows on a sphere (to be published in *Atmósfera* **14**, No. 2, 2001).

23. E. C. Neven, Linear stability of modons on a sphere (to be published in *J. Atmos. Sci*, 2001).

24. E. C. Neven, The power method applied to normal mode analysis of coherent vortex structures on a sphere (submitted to *J. Comp. Phys*, 2000).

PART III

NUMERICAL AND COMPUTATIONAL ASPECTS

DISTRIBUTED PARALLEL SIMULATION OF SURFACE TENSION DRIVEN VISCOUS FLOW AND TRANSPORT PROCESSES

G. F. CAREY, R. MCLAY, W. BARTH, S. SWIFT AND B. KIRK

CFDLab, TICAM
The University of Texas At Austin
Austin, Texas 78712
E-mail: carey@cfdlab.ae.utexas.edu

A domain decomposition approach and finite element formulation are developed for parallel distributed simulation of surface tension driven viscous flow and transport processes. The scheme is implemented in a finite element code MGFLO and is used to study performance on CRAY T3E and SGI Origin 2000 and 3000 parallel supercomputers as well as several PC clusters. The parallel algorithm implementation is briefly discussed, and scaled speedup studies are presented. Representative simulation results for surfactant and thermocapillary driven surface tension flows are also presented.

1 Introduction

The present work concerns our parallel distributed simulation studies for coupled fluid flow and transport simulation. The viscous flow and transport processes considered arise in several application areas such as chemical processing, or semiconductor process technology, as well as in natural phenomena such as volcanic lava flows[1]. Of particular interest are applications where surface tension effects are significant[2-4]. These, in turn, can be strongly influenced by the dependence of surface tension on temperature and chemical concentration in the surface layer. Such effects are important, for instance, in thin films, microscale flows in physical and biological systems, and in microgravity flows.

In the following sections we briefly describe the problem class, discuss the parallel algorithms and implementation, and give performance comparison studies on CRAY T3E and SGI Origin supercomputer systems as well as on PC clusters.

2 Formulation

The class of problems considered involves coupled flow of a viscous incompressible fluid with heat transfer and chemical species transport. Buoyancy is included by means of the Boussinesq approximation as a temperature depen-

144

dent body force term in the momentum equations. The velocity field enters the convective term in the heat transfer (energy) equation and in the species transport equations. The Navier–Stokes equations for viscous flow of an incompressible fluid may then be written as

$$\rho\left(\frac{\partial \boldsymbol{u}}{\partial t} + \boldsymbol{u} \cdot \nabla \boldsymbol{u}\right) + \nabla \cdot \boldsymbol{\tau} = \boldsymbol{f} + \beta(T - T^*)\boldsymbol{g} \tag{1}$$

$$\nabla \cdot \boldsymbol{u} = 0 \tag{2}$$

where $\boldsymbol{u}$ is the velocity field, $\boldsymbol{\tau}$ is the stress tensor (specified by the Stokes hypothesis for a Newtonian fluid), $\boldsymbol{f}$ is an applied body force, $\boldsymbol{g}$ is the gravity vector, T^* is the reference temperature, T is the fluid temperature, and β is the coefficient of thermal expansion of the fluid. At the solid wall boundaries, the no-slip condition applies so that $\boldsymbol{u} = \boldsymbol{u}_w$, where $\boldsymbol{u}_w$ is the specified wall boundary velocity.

The heat equation is given by

$$\rho c_p \left(\frac{\partial T}{\partial t} + \boldsymbol{u} \cdot \nabla T\right) - \nabla \cdot (k\nabla T) = Q \tag{3}$$

where k is the thermal conductivity of the fluid, ρ is the density, c_p is the heat capacity, and Q is a heat source term. Temperature, flux, or mixed thermal boundary conditions may be applied. A similar equation results for transport of a species, and is given by

$$\frac{\partial c}{\partial t} + \boldsymbol{u} \cdot \nabla c - \nabla \cdot (\mathcal{D}\nabla c) = r(c, T) \tag{4}$$

where c is species concentration, $\mathcal{D}$ is diffusivity, and r is the reaction term. Concentration, mass flux or mixed chemical boundary conditions may be applied.

Surface tension effects enter as an applied shear stress which is dependent on the surface temperature and species concentration gradients. For example, on a horizontal free surface, the tangential shear stress component τ_{zx} is given by

$$\tau_{zx} = \frac{\partial \gamma}{\partial x} = \frac{\partial \gamma}{\partial T}\frac{\partial T}{\partial x} + \frac{\partial \gamma}{\partial c}\frac{\partial c}{\partial x} \tag{5}$$

with a similar expression for τ_{zy}, where $\gamma(T, c)$ is the surface tension, T is temperature, and c is the chemical species concentration.

For brevity we will not give the details of the finite element formulation and analysis. It suffices here to remark that we are solving several PDE systems for coupled viscous flow, heat transfer, and reactive chemical species transport in 3D using hexahedral finite elements in an extended primitive variable formulation. A key aspect of the present work is the fact that the driving force in the class of applications considered later is due to the spatial dependence of surface tension on temperature and species concentration in the surface layer. Further details are given in Carey *et al.*[5] Both steady-state and transient applications are considered.

The algorithm requires repeated system solutions within each time step. These systems are nonlinear, sparse and nonsymmetric, but the asymmetry in the applications involving surface tension driven flow is not strong. In the present work we solve the respective systems in parallel over subdomains using Newton iteration with biconjugate gradient iteration (BCG) and diagonal preconditioning for the Jacobian systems[5,6]. Investigations with more sophisticated nested (one-way/cascadic) multigrid and incomplete LU factorization are in progress.

3 Implementation

Parallelism is achieved by partitioning the domain and grid into subdomains comprised of elements. Within each subdomain an element-by-element approach is applied. Communication between processors involves nodes shared on common subdomain boundary segments. Element calculations are made local to each processor and concurrently. (Element matrices are not assembled to subdomain or global matrices.) When the global mesh is partitioned across processors a "send list" of nodes is also set up. These are nodes on the subdomain boundaries and hence are shared by neighboring processors. Each processor then has a send list which puts values in the send buffer and reads from the receive buffer in the same order.

The Krylov subspace BCG solution requires repeated global matrix vector products and vector dot products. For the matrix–vector products, we first initiate computations on the subdomain "boundary strips" and communicate these values to the neighboring processors. During this communication step, the local matrix–vector contributions on the remaining interior elements of each subdomain are carried out. A barrier is inserted at the end of this interior computation so that calculation cannot proceed until remaining communication (if any) is completed. The communicated border values are then accumulated into the local vectors. This results in a correct local processor extraction of the global vector (with duplication of values for nodes shared by neighboring processors). Dot products require a global accumulation of subdomain contributions.

The program is designed for fast scalable parallel computation of steady and transient solutions to coupled incompressible viscous flow with heat and mass transfer. Parallel efficiency is achieved by careful implementation of MPI and customized communication software (such as SHMEM on the T3E) using the domain decomposition strategy mentioned above. Since most of the computation time is taken solving the associated large sparse systems, major effort has been devoted to optimizing the solver.

The input files are described by a simple minilanguage so that the code acts on them as an interpreter. The program has no set script of tasks it performs after reading in the data set. Instead, it performs the steps in the order that the user requests them. This provides great flexibility in how the code is used. The operational statements of the code are written in C, but the actual programming style and implementation is carried out using a higher-level programming tool called **noweb**[7,8]. This has two main features: (1) It permits direct in-line documentation of the code using LaTeX so that to a significant degree, the code can be internally documented (with glossary, index and cross-referencing) which encourages good programming practice; (2) It encourages a structured programming style that has some of the attributes of object-oriented programming independent of the operational language (C or Fortran, etc.).

4   CRAY T3E and SGI Origin parallel supercomputers

Results from scaled speedup studies performed on T3E, Origin 2000 and Origin 3000 systems are shown in Fig. 1. The T3E used consisted of 512 Alpha EV5 processors running at 300 MHz with a 96 kB secondary cache for each processor. Data for the Origin 3000 were taken from a machine with 256 MIPS R12000 processors each of which ran at 400 MHz and had an 8 MB secondary cache. The Origin 2000 results were obtained on two different machines. Both contained 512 MIPS R12000 processors with an 8 MB secondary cache for each processor. The difference was that one system consisted of all 400 MHz processors and the other, 300 MHz. These results include data from a new implementation of the model on the Origin systems using the one-sided SHMEM communication paradigm in addition to the two-sided MPI paradigm.

The performance of the MGFLO model is primarily dominated by the memory bandwidth of the computer system[9]. This can be seen in Fig. 1 when comparing results from the Origin systems and the T3E using SHMEM-based communications. Sustained memory bandwidth on the Origin 2000 system is approximately 400 MB/s per node as measured in the STREAMS benchmark. In that architecture, a node consists of 2 processors. So if both processors on a node are in use, as in this study, the sustained memory bandwidth for each processor is 200 MB/s. With the Origin 3000, the sustained memory bandwidth per node increases to 1.6 GB/s, but a node consists of 4 processors. Thus, sustained memory bandwidth on a per processor basis is 400 MB/s. The STREAMS benchmark on the T3E measures approximately 600 MB/s of memory bandwidth per processor.

The data show the model having the highest performance on the T3E, followed by the Origin 3000 and then by the Origin 2000. This conclusion is further verified by the performance measured on Origin 2000 systems. On this architecture the performance of the model appears to be independent of CPU clock speed. This is attributed to memory bandwidth limitations on the Origin 2000. Memory bandwidth is not related to processor chip speed, but is constant within a computer series. Thus, the performance of the two Origin 2000 systems are seen to be nearly identical. The differences are considered to be noise due to small variations in execution time from run to run.

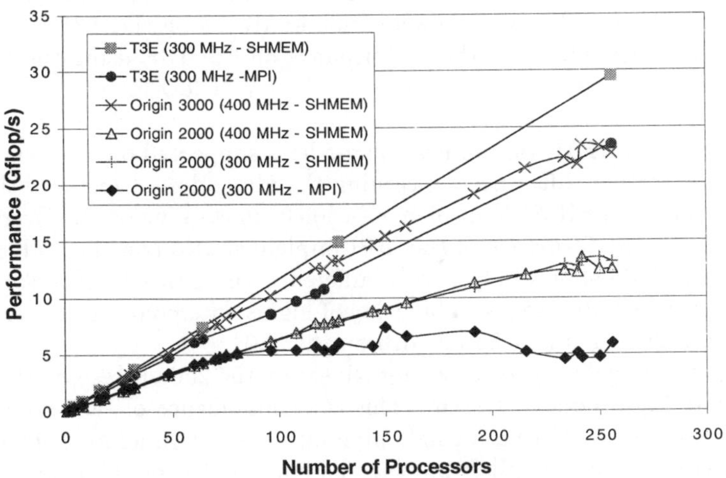

Figure 1. Parallel speedup scaling on the CRAY T3E, the SGI Origin 2000 and the SGI Origin 3000

Comparing the performance of the Origin 2000 (at either 300 MHz or 400 MHz processor speed) to the Origin 3000, the Origin 3000 improves the model performance by a factor of 1.68. Since the memory bandwidth increases by a factor or 2.0, overall performance might be expected to increase by a similar factor of 2.0. However, the bandwidth from secondary cache to the processor does not increase by a factor of 2.0 from the Origin 2000 to the 3000. Thus, the portion of the model run involving secondary cached data sees no reduction in execution time and no performance gain from one architecture to the other.

A final performance issue is related to communication technology. On the Origin systems, the model has recently been implemented using the SHMEM paradigm for communications. This greatly improved scalability on those machines. Comparing data on the 300 MHz based Origin 2000, it is seen that SHMEM and MPI give identical performance on up to 80 processors. On more than 80 processors, the MPI scalability levels off and starts to show a great deal of variability from one processor count to the next and from one run to the next. Use of SHMEM greatly increases scalability and reduces the variation in performance. This is due to the lower overhead of the one-sided SHMEM communication versus the two-sided MPI communication. On the T3E, MPI shows scalability similar to that of SHMEM, but has lower performance at all processor counts. This is because SHMEM takes direct advantage of lower latency communication hardware on the T3E.

It is not surprising that the performance on the T3E using the SHMEM communications protocol is the highest. This is a low-latency, single sided communications language whereas MPI has greater overhead and is two sided. Although the SHMEM communication model is supported on the Origin 2000, there is one further difference concerning the allocation of memory that currently prevents our SHMEM implementation from working on the Origin 2000. Comparing the performance of the MPI implementation on the T3E and 300 MHz Origin 2000 shows only a 20% difference through 64 processors. This difference is attributed to a combination of higher interconnect network bandwidth and higher memory bandwidth on the T3E.

Second, the scaled speedup is more linear on the 256 processor Origin 2000 than on the 128 processor system. This is a consequence of each node having two processors, but sharing a single pipe into the local memory. The subdomain partitioning and explicit message passing of the model insures that no processor directly accesses memory on another node. Through 64 and 128 processors on the 128 and 256 processor machines, a dedicated environment was set up specifically to use only one processor on each node. This guaranteed that each processor attained the maximum memory bandwidth possible. The strong degradation in performance seen on the smaller system when running on more than 64 processors is due to at least some of the processes running on the same node, effectively reducing the memory bandwidth.

Third, the scaling performance drops off again past 120 processors on the smaller system. This degradation in performance is due to the overhead of

IRIX on the system. As the test case is scaled up, system load balancing shifts the various IRIX processes to unused nodes and processors. Hence, as we scale up to near the system size, there are no idle processors to run IRIX and some process swapping is done between the model and the OS. The model requires barriers at each iterate step, so the entire model typically runs no faster than the slowest process.

Another feature is that for a 54% clock speed gain (195 MHz to 300 MHz), the model speedup is only 20% (as mentioned above). This is attributed to the model having little cache reuse and being somewhat memory bandwidth limited. The faster system has a larger cache but the same memory bandwidth. Similar behavior is seen comparing T3Es with various processor speeds (300 and 450 MHz), but the same cache and memory bandwidth in each case.

Additionally, it is noted that the scaled speedup efficiency for the Origin is different when using only one processor on a node compared to using both processors. For 64 or fewer processors, the scaled speedup efficiency is 88%. This means that when increasing the number of processors by a factor, n, the overall performance will increase by only $0.88n$. For more than 64 processors, the efficiency is only 79%. This is true through about 120 processors on the studied system. No explanation is apparent at this time.

The steady state flow pattern for surfactant driven flow on a slice through a cube domain is indicated in Fig. 2. The surfactant enters via a Gaussian flux on the top free surface. The five remaining walls are specified to have zero concentration Dirichlet-type boundary condition. The surface tension on the top free surface is a function of surfactant concentration gradient and causes a large convective cell to form.

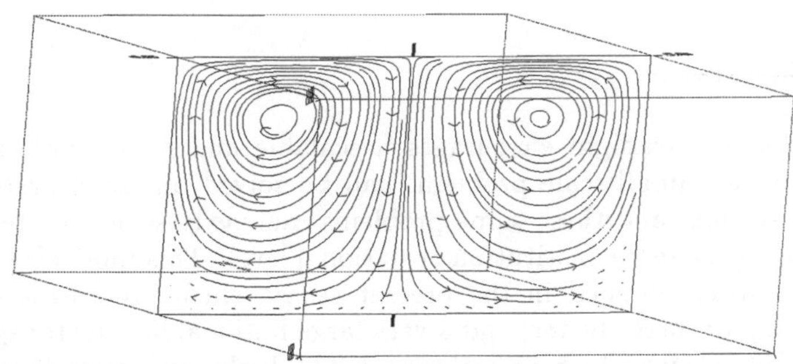

Figure 2. Steady state surfactant-driven Rayleigh–Benard–Marangoni flow in a cube.

Results for a reacting flow case are shown in Figs. 3 and 4. Figure 3 shows the flux distribution of two separate reactant species on the top boundary of a cube domain. The five remaining walls are specified to have zero species concentration. These two species react to form a third (whose concentration is proportional to the product of the two reactants). The solution was marched in time from an initial state (with no reactants or product in the domain) until it reached a steady state. Figure 4 shows the steady-state concentration on parallel slices through the domain for the three species. The figures indicate that for this case diffusion is the dominant process.

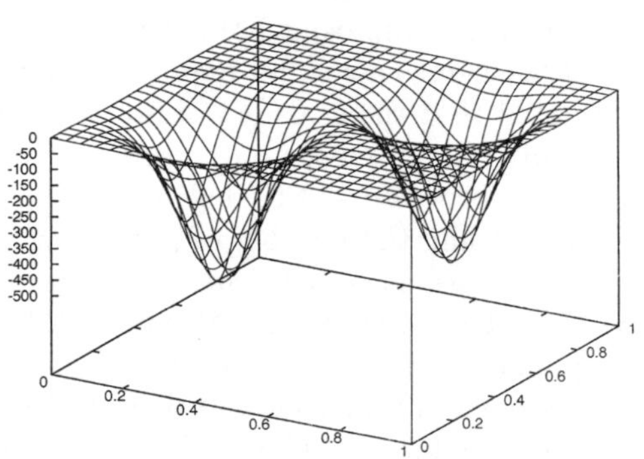

Figure 3. Flux boundary condition for two reactant chemical species

5 PC clusters

The PC cluster paradigm encourages a scalable and expandable approach to applications. Small-scale problems can be solved on small systems in a laboratory setting, and these same problems can easily scale to intermediate and large systems at the institution or national level. This model is consistent with the strategy pursued in the present study which involves a small lab cluster, a larger center cluster, and a very large national lab cluster system. It is also consistent with a hierarchical model of scalable computing that includes tightly integrated distributed large-scale supercomputers like the T3E or SGI Origin.

Three cluster systems of increasing size and complexity were investigated in this study:

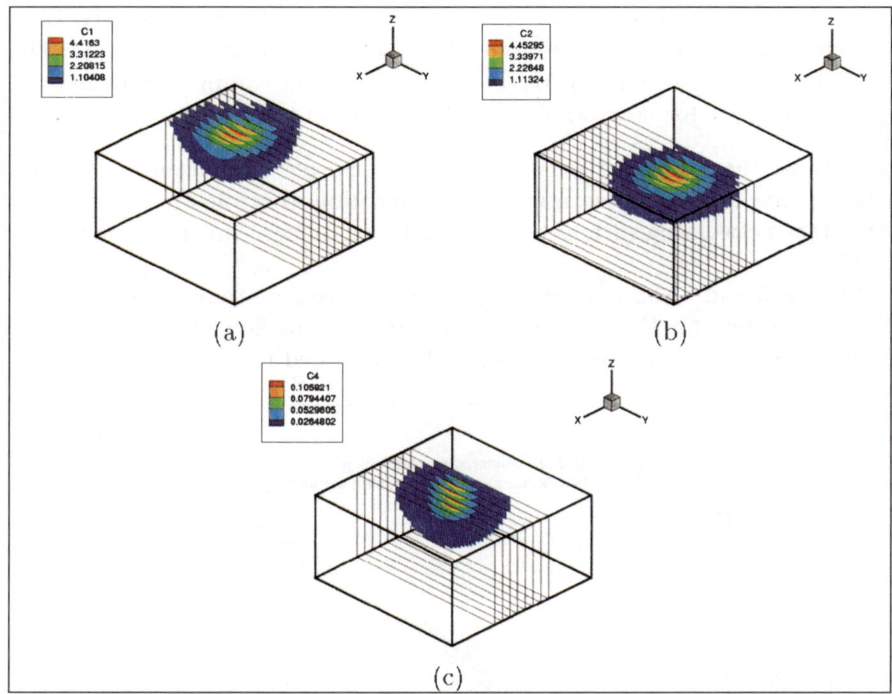

Figure 4. Steady-state concentrations of (a) species 1, (b) species 2, and (c) reaction product

1. The CFDLab cluster, built under a grant from Intel, is comprised of 16 Intel Pentium II 266 MHz processors used for the investigation of laboratory computing on workstation clusters. Each node is equipped with a single processor and 128 MB of RAM. The compute nodes are interconnected with a single crossbar switch running 100 Mbit FastEthernet. The cluster runs the RedHat 6.1 Linux operating system, uses GNU compilers, and communicates via MPI[10].

2. The TICAM (Texas Institute for Computational and Applied Mathematics) cluster, "Longhorn," was assembled from 64 Pentium II 300 MHz processors, each with 512 MB of RAM and connected via Myrinet and FastEthernet. Like the CFDLab cluster, the individual nodes run RedHat Linux and communicate via MPI[11].

3. "Alaska," at Sandia National Laboratory, consists of up to 592 Alpha

workstations with processor clock speeds of 500 MHz and 256 MB RAM per node (230 of which were available at the time of this study). The nodes are connected via Myrinet for use during computations and via FastEthernet for user/administrator interaction[12].

Because these workstations are part of a cluster built from commodity parts, we have neither a customized communications controller nor high-speed proprietary networking such as those available on modern supercomputers like the T3E.

Figures 5 and 6 show the results of scaled speedup studies performed on the three clusters for the 3D simulation of the viscous flow and heat transfer problem. The scalability on a CRAY T3E is included for reference in Fig. 6.

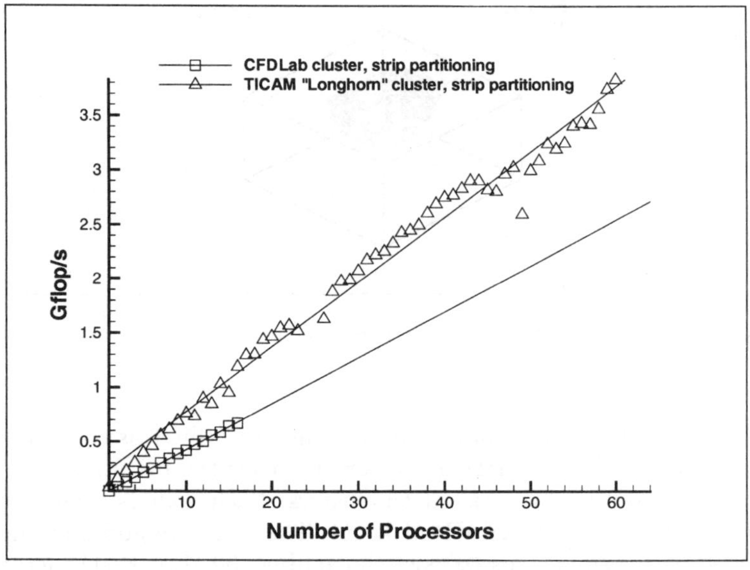

Figure 5. Parallel speedup scaling on Beowulf clusters

Scalability studies were performed on two domains of a different aspect ratio: (1) the "rod" benchmark divides a high-aspect ratio rectangular rod-like solid domain into a number of cubes using a number of parallel slices. This partitioning method creates equally spaced subdomains, which are in turn assigned to a processor. This partitioning method requires each processor to share information with no more than two other processors during the system solution. Thus, this method produces superior scalability. (2) The

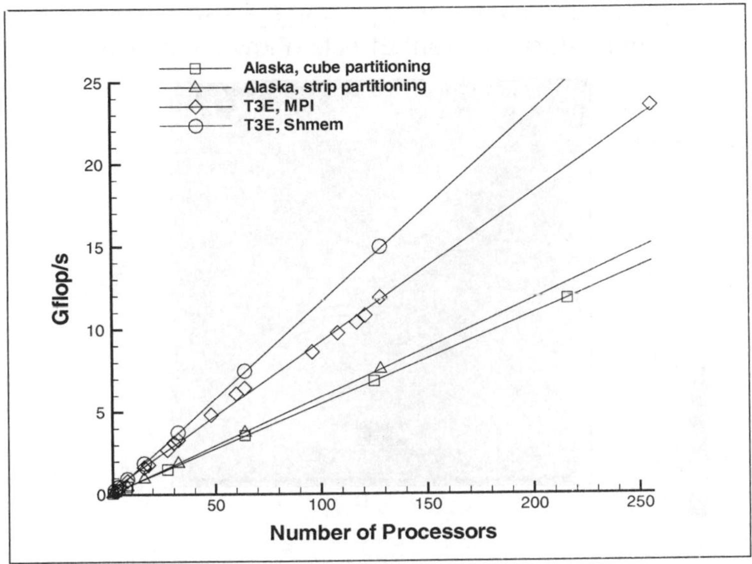

Figure 6. Parallel speedup scaling on Alaska and a CRAY T3E

"cube" benchmark, on the other hand, divides a cube domain with slices in the three orthogonal coordinate directions. This partitioning strategy maximizes the number of face, edge and vertex processor neighbors and therefore the amount of communication required during the system solution. These two cases provide an indication of the range of communication performance of a cluster and algorithm implementation.

The results presented in this section are for a steady-state simulation for thermocapillary convection driven flow in a fluid cell[1,3,4]. All side walls have a no slip condition, and a no-flux thermal condition is imposed on the front and rear faces of the cube shaped computational domain. A constant, $u = 1$, velocity condition is also imposed on the lower surface of the domain. The temperature on the left and right side walls is specified as shown in the figures and an insulating boundary condition is applied on the bottom surface.

The top boundary is a horizontal free surface with a mixed thermal boundary condition having reference temperature of 0.5 and a convective heat transfer coefficient of 1.0. This models the presence of a cooling gas above the domain. A 3D steady-state simulation is made and 2D section plots are shown in Fig. 7.

Figure 7 shows three distinct vortices in the resulting flow field. The

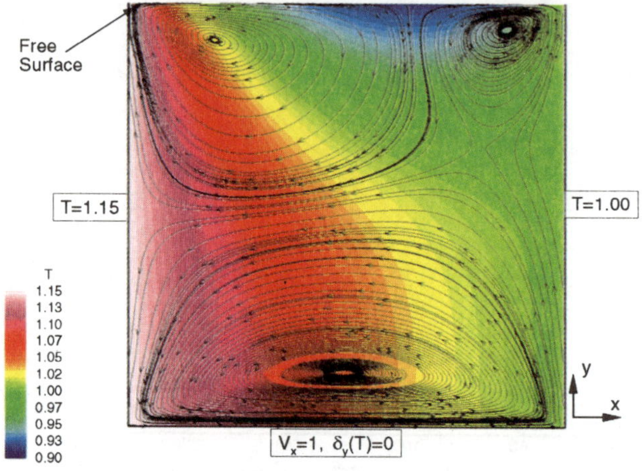

Figure 7. Driven cavity flow and temperature fields

bottom vortex is associated with the velocity imposed on the bottom boundary of the domain. The two vortices at the top of the domain are associated with the surface tension effects. The surface tension tends to draw the fluid on the top surface from hot regions to cooler ones. Since this surface is cooling to a temperature less than that of both walls, the surface tension induces this counter-rotating vortex pair. The left of the two vortices appears stronger because it has formed over a larger temperature gradient than the left vortex. This is supported by the contours in the figure (which show the temperature field for this case).

6 Conclusions

This investigation describes progress on the development of methodology, algorithms and software for large scale multiphysics flow and transport simulation. The focus in the present work is on nonlinear problems where surface tension effects are important. The code MGFLO utilizes a gradient iterative solver and simple preconditioning. Consequently, it scales very well on both traditional parallel supercomputers and to a large number of PC cluster configurations. The code scaling on PC clusters was found to be slightly dependent on the partitioning method employed, however, due to the overhead imposed by interprocessor communication. On traditional parallel supercomputers the code scaling was influenced predominantly by memory bandwidth and, at the full size of the machine, competition with the operating system for system resources. Improved preconditioners are needed for efficient iterative performance and this is under current development. We have made preliminary studies with one-way multigrid for the steady-state problem and results are encouraging.

Acknowledgments

This research has been supported by NASA ESS Grand Challenge Project NCCSS-154. The Beowulf cluster used in work has been supplied via an equipment grant from Intel.

References

1. E. L. Koschmieder, *Bénard Cells and Taylor Vortices* (Cambridge University Press, New York, 1993).
2. C. Cuvelier and J. M. Dreissen, *J. Mech.* **169**, 1–26 (1986).
3. Y. Kamotani, S. Ostrach and A. Pline, *J. Heat Transf.* **117**, 611–617 (1995).
4. J. R. A. Pearson, *J. Fluid Mech.* **4**, 489–500 (1958).
5. G. F. Carey, R. McLay, G. Bicken, B. Barth, *IJNMF* **31**, 37–52 (1999).
6. G. F. Carey, R. McLay, C. Harle, S. Swift and W. Spotz, in *High Performance Computing '98 Proceedings* (Boston, MA, April 1998).
7. N. Ramsey, *IEEE Software* 97–105 (1994).
8. N. Ramsey, Noweb Homepage, `http://www.eecs.harvard.edu/~nr/noweb/`, 1999.
9. S. Swift, R. McLay and G. F. Carey, in *High Performance Computing 2000 Proceedings* (Washington, DC, April 2000).
10. CFD Lab, The University of Texas at Austin, `http://www.cfdlab.ae.utexas.edu/intel_beowulf/`
11. Longhorn: Center for Subsurface Modeling, 64 Node Beowulf System, `http://www.ticam.utexas.edu/CSM/longhorn.html`
12. Computational Plant, Sandia National Laboratory, `http://www.cs.sandia.gov/cplant`

CONSIDERATIONS FOR SCALABLE CFD ON THE SGI CCNUMA ARCHITECTURE

M. D. KREMENETSKY, S. A. POSEY

*Silicon Graphics, Inc., 1600 Amphitheatre Pkwy.,
Mountain View, CA 94043-1351, USA
E-mail: mdk@sgi.com, sposey@sgi.com*

T. L. TYSINGER

*Fluent, Inc., 10 Cavendish Court, Lebanon, NH, USA
E-mail: tlt@fluent.com*

Recent breakthroughs in CFD technology have led to a high degree of parallel scaling of CFD simulations in a distributed shared-memory environment. A variety of research and commercial CFD software demonstrate scalable levels that are linear beyond 256 processors on the ccNUMA system architecture developed by SGI. Parallel algorithms typical of contemporary CFD software offer efficient strategies that overcome bottlenecks for moderate levels of parallelism. However, to achieve efficient parallelism on 100's of processors, additional consideration must be given to topics such as performance of system software and awareness of communication architectures. This paper examines the requirements for highly scalable CFD simulations on an assortment of industrial application examples.

1 Introduction

Traditional industries such as automotive, aerospace and power generation are challenged with an increasing need to reduce development cycles, while satisfying global regulations on safety, environmental impact and fuel efficiency. They must also appeal to demands for high quality, well-designed products in a competitive business environment. Continuing technology advances in computer-aided engineering (CAE) simulation provide industry with a design aid that is a relevant step towards achieving these goals.

Historically, CAE simulation provided limited value as an influence on industrial design owing to excessive modeling and solution times that could not meet conventional development schedules. During the 1980's vector architectures offered greatly improved CAE simulation turnaround, but at a very high cost. RISC computing introduced in the 1990's narrowed this gap of cost–performance, however bus-based shared-memory parallel systems did not scale sufficiently beyond 8 processors.

Recent advancements in parallel computing have demonstrated that vector-level performance can be easily exceeded with proper implementation of parallel CAE algorithms for distributed shared-memory systems like the SGI 2800. Perhaps even more appealing is that the increased performance is offered at a fraction of the cost. These trends have influenced recent increased investments by users of CAE technology throughout a range of industries.

The automotive industry in particular has made substantial investments over the past three years in scalable systems and parallel CAE software. It is estimated that during 1999 alone, total GFlop/s computational capacity within the three Detroit automotive OEM's increased more than 2-fold. This is in large part due to efficient parallel implementations of commercial CAE software for applications such as crashworthiness simulation, analysis of noise, vibration and harshness (NVH), and computational fluid dynamics (CFD) simulation. During a one year period, a total of 740 processors in SGI 2800 servers including two systems with 128 processors each, were deployed within the Detroit Big 3, representing about a 375 GFlop/s increase to existing computational capacity.

A system architecture's ability to achieve high parallel efficiency becomes increasingly important as algorithms for CAE software applications are developed towards such capability. From a hardware and software algorithm perspective, there are roughly three types of CAE simulation "behavior" to consider: implicit and explicit finite element analysis (FEA) for structural mechanics, and CFD for fluid mechanics. Each have their inherent complexities with regards to efficient parallel scaling, depending upon the parallel scheme of choice.

Most commercial CAE software employ a distributed-memory parallel (DMP) implementation based on domain decomposition methods. This method divides the solution domain into multiple partitions of roughly equal size in terms of required computational work. Each partition is solved on an independent processor, with information transferred between partitions through explicit message passing software (usually MPI) in order to maintain the coherency of the global solution.

Other choices for efficient parallel methods are shared-memory parallel (SMP) coarse grain, and hybrid parallel schemes that combine DMP and SMP within a single simulation. Hybrid parallel schemes, which are becoming increasing popular for applications that contain a mix of Eulerian and Lagrangian mechanics such as combustion, are particularly well suited to the SGI 2800 distributed shared-memory architecture.

The distributed shared-memory SGI 2800 system is based upon a cache-coherent nonuniform memory access (ccNUMA) architecture[1]. Memory is physically distributed but appears logically as a shared resource to the user. Motivation for ccNUMA evolved at SGI as conventional shared-bus architectures like that of the SGI Challenge would exhibit high-latency bottlenecks as the number of processors increased within a single system. During this same time, noncoherent distributed-memory architectures emerged, but application development for most designs was considered too difficult for commercial success.

The SGI ccNUMA implementation distributes memory to individual processors through a nonblocking interconnect design, in order to reduce latencies that inhibit high bandwidth and scalability. At the same time, a unique directory based cache-coherence provides a memory resource that is globally addressable by the user, in order to simplify programming tasks. A single image SGI 2800 system offers up to 512 processors and can expand to 1 Tbyte of memory, which is the largest SMP system currently available in industry.

2　Industrial application examples

Three industrial-size CFD application examples are provided that demonstrate the requirements and recent achievements towards highly scalable CFD. Each example presents models from industry and represent some of the largest currently in practice. These industry examples highlight applications from commercial and research CFD.

Algorithms and applications for CFD have led the advance in parallel scalability. CFD models are growing in size such that the ratio of work to overhead is rapidly increasing. Also, fluid domains have a natural load-balance advantage over structural domains owing to uniformity of fluid properties — typically air or water, such that model entities for CFD typically offer uniform computational expense from one to the next.

Structures often contain a mix of materials and finite elements that can exhibit substantial variations in computational expense, which creates load-balance complexities. The ability to efficiently scale to a large number of processors is highly sensitive to load balance quality. The examples illustrate these differing characteristics between the two disciplines.

2.1　Commercial CFD — FLUENT

Commercial CFD software such as FLUENT from Fluent Inc.[2] is among the most efficient parallel software in industry. Fluent in particular has provided leadership in the recent trend of CFD simulations reaching essential higher resolutions. FLUENT is a multipurpose, unstructured-mesh CFD solver that employs an algebraic multigrid (AMG) scheme for solution of the 3D, time-dependent Navier–Stokes equations.

The parallel implementation of FLUENT can accommodate homogenous compute environments such as SMP systems or clusters, as well as a heterogeneous network of various types of systems. A domain decomposition method with various graph partitioning schemes is used as the parallel strategy. The parallel programming model used is DMP with MPI as the message passing software.

For classic DMP implementations, several factors can inhibit parallel scalability. CFD solver algorithms affect the frequency and amount of information that must be shared across partitions. Similarly, efficient planning of what data to share (the message content), and when to do so, is important. Quality domain decomposition schemes are critical since they affect load balancing, and determine the size of partition boundaries and consequent message passing requirements.

Efficient algorithmic and partitioning strategies are provided in FLUENT that overcome the scaling issues noted. Linear scaling to about 16 processors is routinely achieved with well-designed parallel CFD software. Beyond that level, efficient scaling can be sensitive to the performance of hardware communication architectures and system software. In particular, the choice and implementation of the system software for message passing between partitions is critical.

Parallel efficiency of the AMG solver is restricted only by hardware and software communication latency of a particular system. FLUENT supports proprietary versions of MPI, some with latency improvements of nearly 3-fold over public domain MPICH. During 1997, SGI released an MPI based on standard MPICH source that was tuned to properly account for performance issues related to the SGI ccNUMA architecture.

The ccNUMA-aware MPI provided by SGI exhibits a latency of about 13 μs on the SGI 2800 system. Further experiments conducted on the SGI 2800 with a one-sided MPI implementation based upon the MPI-2 standard, show latency in the range of 3 μs. These results suggest that further improvements of FLUENT parallel efficiency are possible in the near future.

A recent investigation on the parallel efficiency of a large automotive case involved an aerodynamics study of a full vehicle on a stationary ground plane. The flow conditions were steady-state and isothermal, and a k-ϵ turbulence model was used to resolve shear effects. The most critical aspect of this model was its size — 29M cells, which exceeds recent automotive aerodynamics modeling by a factor of 3.

The system used for this aerodynamics study was a SGI 2800 with 256 MIPS processors[3] at 195 MHz, and 33 Gbytes of memory. A segregated solver strategy in FLUENT 5.1.1 was used for solution of the 29M cell model, which

Table 1. FLUENT Performance for 29M Cell Automotive Aerodynamics Example

Number of CPUs	Seconds/Iteration	Parallel Speedup
10	381	1.0
30	99	3.9
60	67	5.7
120	29	13.1
240	18	21.2

required 28 Gbytes of memory. As with performance evaluations on other large automotive examples, this case exhibited a remarkably high degree of parallel efficiency for FLUENT. The results are summarized in Table 1.

Parallel performance results from 10 to 240 processors produced a 21-fold speedup, close to ideal which occurs at 24-fold. Solution on 10 processors required 381 seconds per iteration and was reduced to just 18 seconds for 240 processors. The speedup from 10 to 120 processors was 13-fold, or greater than ideal at 12-fold. This "superlinear" behavior can occur when domain partitions are a size where they become local-memory resident. Local memory is 256 Mbytes and the domain contains 120 partitions of roughly equal size, each about 233 Mbytes.

The study measured performance starting with 10 processors rather than one, in order to reduce the artificial scaling benefits observed for large remote-memory access times. With 10 processors, some of the threads are distributed to nodes with data in local memory. This reduces the overall number of remote-memory hops since 10 threads require less hops than a single thread. Using the 10 processor result as a reference time is more realistic for a scalability study of this model size.

This achievement is significant for the automotive industry since aerodynamic characteristics are important during early development efforts that fix the vehicle shape. CFD simulation is generally not influential at this early stage. Instead, conventional aerodynamic evaluation of vehicles rely primarily on expensive wind tunnels since an equivalent CFD simulation would require among other things, models with resolution levels near the 29M cell case.

For its part, the performance limitations of large model turnaround for vehicle aerodynamics has been addressed with this study. Additional CFD modeling features, mostly related to turbulence, are required before CFD simulation can effectively replace wind tunnels. Still, CFD capability has reached an effective level of accuracy and economics today as an established experimentation alternative, or at least a complement, for a wide range of applications.

In another FLUENT example involving combustion simulation for an industrial burner, the parallel benefits of hybrid SMP and DMP methods are illustrated. The model contains 155K cells, and in addition to resolution of momentum and energy distribution, simulations of this class require accurate modeling of spray and chemical combustion phenomena. Combustion for this model includes chemical reactions for 6 species, and tracking of 500 disperse phase particles.

Prior to the release of FLUENT 5.1.1, models of this type would exhibit very poor parallel scalability since the highly parallel Eulerian continuum gas phase was coupled with the serial Lagrangian disperse phase. Implementation by FLUENT of an SMP particle decomposition with mfork threads now permits models such as these to achieve near-ideal parallel speedup. For the 155K cell burner example, the hybrid scheme shows more than a 7-fold speedup on 8 processors.

2.2 Research CFD — OVERFLOW

Research CFD software is typically developed for special purpose modeling. OVERFLOW from NASA Ames[4], is a structured over-set grid CFD code that is applied by both research and industry for investigation of aircraft aerodynamics. It solves the 3D time-dependent Reynolds-Averaged Navier–Stokes equations. Unlike most commercial CFD software that use a DMP approach, OVERFLOW uses an SMP approach that is multilevel parallel (MLP), meaning both coarse and fine-grain SMP are employed within a single simulation.

The MLP approach uses fine-grained, compiler-generated parallelism at the loop level, and at the same time performs parallel work at a coarser level using standard Unix fork system calls. During execution, load balance adjustments are fully automatic and dynamic in time. Global data is shared among the forked processes through standard shared-memory arenas. This eliminates the need to design code for synchronization and communication as would be required for DMP, and as such greatly simplifies code development.

The shared-memory MLP approach has performance advantages over DMP since it offers coarse-level parallelism without the need for message passing. Since DMP requires an additional software interface such as MPI, latencies and other overhead are imposed that affect scalability. These messaging subroutines are used to communicate boundary values and other data between partitions as the solution progresses, which is not required for MLP since data is shared globally.

162

Table 2. OVERFLOW Performance for 35M Cell Aircraft Aerodynamics Example

SYSTEM	Processors	GFlop/s
CRAY C90	16	4.6
Origin2000	256 × R12000/250Mhz	20.1
Origin2000	512 × R12000/250Mhz	37.0
Origin2000	512 × R12000/300Mhz	60.0

Beginning in 1997, NASA Ames began deployment of several SGI 2800 systems, the largest of which is a 512 processor single system image (SSI) and currently the largest SMP system in industry. Several large aerodynamic simulations have been conducted with OVERFLOW-MLP up to the maximum 512 processors. One example is a model of a full aircraft configured for landing that contains 35M grid points distributed over 160 grid zones. At the time, this model was considered among the largest ever attempted at NASA Ames.

Performance results for OVERFLOW-MLP 1.8 on the 35M grid-point model, which are summarized in Table 2, were obtained for three NASA systems: CRAY C90/16, SGI 2800/256 (250 MHz), and SGI 2800/512 (300 MHz). The level of achieved GFlop/s during the simulation was measured for each system and the C90 was measured at 4.6, the 2800/256 at 20.1, and the 2800/512 at 60.0 — a performance level that required only 2.5 seconds per time step. Parallel performance for both SGI 2800 systems achieved linear scaling to their maximum number of processors.

The 35M grid-point OVERFLOW study demonstrates several significant computing advances. Most important is the practicality for model resolutions as high as 100M grid points — a level considered necessary for "virtual wind tunnel" simulations. Another advance is the cost–performance improvements of a new generation of parallel architectures. NASA reports the cost of the SGI 2800/512 to be 2.6-fold less than the CRAY C90. With a performance advantage of 13-fold, this provides the 2800/512 system with a 33-fold cost–performance advantage.

3 Future directions

In the wake of recent breakthroughs for scalable CAE simulation, research and industry will continue to increase their investments in CAE technology as a product and process design aid. The motivation is simply a matter of economic benefits and improved quality that scalable CAE brings to the development process. Efficient turnaround of CAE simulations means increased modeling resolution and more comprehensive evaluation during. early development stages.

Advancements will continue well into the new decade for improved CAE scalability as emerging algorithm developments and new hardware architectures lead a path towards enhanced CAE methodologies. These enhancements will encourage an increase in CFD modeling for transient flow conditions, widespread implementation of probabilistic structural mechanics, and production capability for multidiscipline fluid and structure coupling, among others.

CFD simulations for transient conditions have reached their potential for industrial application. Important simulation capability for transient flows include applications of automotive powertrain in-cylinder combustion, vehicle aerodynamics with rotating wheels and moving ground plane, commercial aircraft aerodynamics for off-design (noncruise) conditions, and a variety of turbomachinery aero-thermal flows for compressor, combustor and turbine designs.

Structural FEA simulations are currently undergoing a historic transition from deterministic to probabilistic. Turnaround for a single FEA analysis is considered small enough that highly parallel stochastic techniques are being applied to better manage design uncertainty of scatter observed in sources such as material properties, test conditions, manufacturing and assembly. The discipline of explicit FEA in particular is suited to benefit from probabilistic techniques.

The high-transient nonlinear modeling of dynamic events with explicit FEA, such as automotive vehicle crash, airbag–occupant interaction and aircraft bird-strike all exhibit substantial parameter scatter. This trend towards stochastic simulation is a relevant step towards an overall trend towards single and multidiscipline optimization. Lately, Monte Carlo stochastic methods have been applied to automotive vehicle design for improved NVH and crashworthiness.

Computer hardware architectures will continue to improve. During 2000, SGI introduced a new generation of ccNUMA systems that provides a 2-fold increase in system bandwidth with a 2-fold decrease in latency. Other performance improvements include a processor clock speed increase by 33% to 400 MHz, and a larger L2 cache. These and other features improve both single processor turnaround and parallel efficiency over the current SGI 2800 ccNUMA system.

4 Conclusions

Effective implementation of highly parallel CFD simulations must consider a number of features such as CFD parallel algorithm design, system software performance issues, and hardware communication architectures. Several examples of commercial and research CFD simulations demonstrate the possibilities for efficient parallel scaling on the SGI ccNUMA system.

Development of increased parallel capability will continue on both software and hardware fronts to enable modeling at increasingly higher resolutions. Indeed, the demonstrated benefits of cost-effective, highly scalable CFD simulation has potential to shift modeling practices within research and industry on a global basis, and continue to further advancements in industrial design applications.

Acknowledgments

The authors wish to acknowledge certain colleagues for their contributions to this paper; in particular, Ayad Jassim of SGI and Jim Taft, consultant to NASA Ames.

References

1. J. Laudon and D. Lenoski, "The SGI Origin ccNUMA Highly Scalable Server," (SGI, Mountain View, CA, 1997).
2. FLUENT 5 User's Guide (Fluent Incorporated, Lebanon, NH, 1998).
3. K. Yeager, "The MIPS R10000 Superscalar Microprocessor," *IEEE Micro* **16**, 28–40 (April 1996).
4. J. Taft, "Performance of the OVERFLOW-MLP CFD Code on the NASA Ames 512 CPU Origin System," *Proceedings of the 2000 HPCC/CAS Workshop* (NASA Ames Research Center, Mountain View, CA, February 2000).

PARALLELIZATION OF A TWO-WAY INTERACTIVE GRID NESTING ATMOSPHERIC MODEL

PATRICK JABOUILLE

*Centre National de Recherches Météorologiques, Météo-France,
42 av G Coriolis, 31057 Toulouse Cedex, FRANCE
E-mail: patrick.jabouille@meteo.fr*

Numerical simulations of the atmospheric motions require the most powerful machines and a high performance computer technology. The French meteorological model MésoNH is able to perform several simultaneous simulations on a nested grid to focus on specific regions described by a higher spatial resolution. This paper describes software and techniques used for implementing the MésoNH model on parallel processor computers especially for the grid nesting part.

1   Introduction

The MésoNH atmospheric simulation system is a joint effort of Centre National de Recherches Météorologiques (CNRM Météo-France) and Laboratoire d'Aérologie (CNRS). It is designed as a research tool for small and mesoscale atmospheric processes. It comprises several elements: a numerical model with a comprehensive physical package, an ensemble of facilities to prepare initial states, postprocessing and graphical tools to visualize the results.

MésoNH is a gridpoint limited-area model. To be able to use a horizontal resolution lower than 5 km, the model is nonhydrostatic, i.e., it uses a complete prognostic equation for the vertical speed. This assumption implies solving an elliptic equation to determine the pressure. The model solves the equations governing the atmospheric state evolution on a computational grid. This grid is nonorthogonal to maintain the vertical coordinate and take into account the topography (more details about the MésoNH equations and discretization can be found in Lafore *et al.*[1] or by visiting our web site: http://www.aero.obs-mip.fr/mesonh/index2.html). In addition to the dynamic part of the flow, the model takes into account many physical parameterizations (turbulence, radiations, surface processes, microphysics, etc.). It allows for the transport and diffusion of passive scalars, to be coupled with a chemical module.

The code is entirely written in Fortran 90 and in a first stage it has been developed to be run on a single-processor computer. In order to improve the spatial resolution and the representation of physical processes, a parallel implementation of the MésoNH model has been recently achieved. This work has been done in collaboration with the CERFACS[a] laboratory. The parallelization of the grid nesting part, which has required a special approach, will be described in greater detail in the paper.

2   Interface routines

The parallelization work began by a precise specification of requirements related to communications between processors to run the model on a parallel machine:

- Decomposition of the model on n processors.

- Parallelization is achieved to be as much as possible transparently to users not aware of parallelization techniques.

- Full compatibility of model running on one processor with the same code.

- Portability to any computer having the MPI library.

- Routines allowing I/O flow along a transparent way for users.

To meet these goals, an interface library named *ComLib* (for Communication Library) has been developed. It contains all routines necessary to parallelize the MésoNH model and is based on the standard library MPI (Message Passing Interface). This package is implemented in Fortran 90 on top of the MPI library. Its main purpose is to provide the MésoNH developers, who are not necessarily expert in parallel computing, with all the required communication capabilities while hiding from them the low level message passing paradigm details. Similar approaches have been developed to parallelize meteorological models: RSL (Runtime System Library) by John Michalakes implemented in MM5, and MSG (Message passing tools for Structured Grid communications) by Andrei Malevsky for MC2 (Canadian model).

The implemented coarse grain parallelism exploits a classical parameterizable 2D decomposition in the x–y direction of the 3D physical domain (the vertical dimension is not decomposed). The automatic partitioning of the 3D domains generated by *ComLib* produces as many nonoverlapping vertical rectangular beams as processors requested by the user. Each beam (hereafter called subdomain) is then assigned to one processor of the parallel target computer. Some overlapping data structures referred to as "halos" in the sequel are required; this enables reusing all of the computational routines existing

---

[a]Centre Européen de Recherche et de Formation Avancée en Calcul Scientifique

in the sequential version of the MésoNH code. Because most of the operators are discretized using a 5 point stencil in the horizontal plane, the "halo" can be just one cell width. However this width may be increased to suppress supplementary communications. If the rectangular decomposition minimizes the data amount of the overlapping area, it could be useful to decompose the domain in stripes along the x axis to promote the vectorization of the numerical computation. It is possible for the user to chose the way of decomposition (stripes or rectangles) and the size of the halo.

The *ComLib* library is based on data structures organized in an object-oriented style, with Fortran 90 user-defined types. There are two top-level types:

- The configuration of all the processors, i.e., the way the domain is split into subdomains corresponding to the processor.

- The communications to be performed by the current processor. They are managed by several lists of zones, one for the send and one for the receive operations for each kind of communication.

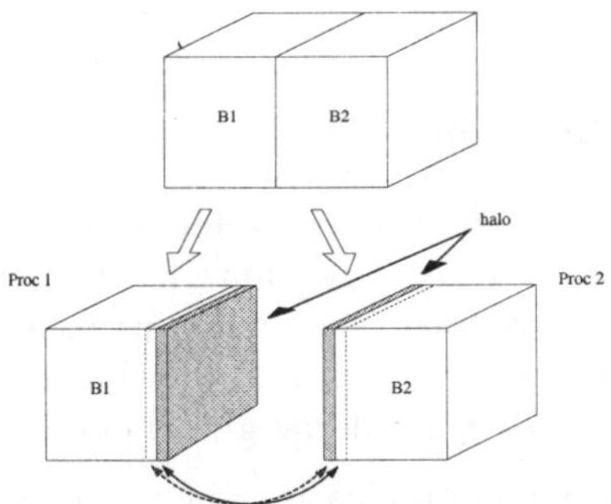

Figure 1. Example of decomposition.

In Fig. 1 we depict an example of decomposition suitable for a parallel execution of MésoNH on a two processor computer. Beam B_1 and B_2 are respectively allocated to processor 1 and 2; the computation on subdomains in B_i will be correct if data in the halo (shadowed area in Figure 1) have been updated correctly prior to the computation. It is up to the user to define the list of fields that should be kept updated in the halo region through specific subroutines to manipulate lists of fields, as well as when this field list should be updated by a call to a *ComLib* routine.

The distribution depicted in Fig. 2(a) does not work to solve the elliptic pressure system because a fast Poisson solver based on Fast Fourier Transform (FFT) computations are used. Two other types of distribution are used, shown in Figs. 2(b) and 2(c). Communication routines have been implemented that move a field between these different decompositions, which makes it possible to perform the FFT for each horizontal direction.

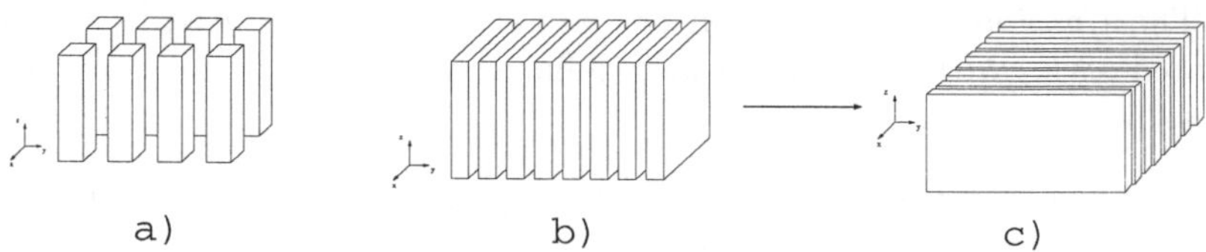

Figure 2. The different distributions used in the model.

The IO routines have been modified in order to carry out at the same time the reading and the distribution of the input fields and the gathering and the writing of the output fields, in a transparent way for the user.

3   MésoNH parallelization

The parallelization of the code has been achieved by modifying some routines in the temporal loop to take into account the data domain decomposition. Three types of modified routines can be distinguished: routines to communicate fields between processors, routines allowing the treatment of physical boundary conditions and routines doing global operations.

- The main processor communications in a time step are located at the end of temporal loop to refresh the halo of all prognostic variables (wind, temperature, humidity, etc.). If only one width of the halo area is used, other supplementary communications for a few fields are required to obtain a correct computation in the inner subdomain (communication of surface fluxes, for example).

- Modifications in some MésoNH routines have been made to distinguish processors located on physical borders. Special *ComLib* functions are used to do this.

- Global operations using the Fortran 90 functions SUM(A) or MIN(A), where A is an array, have been replaced by equivalent parallel *ComLib* functions involving processor communications.

It can be noticed that the modifications in the Fortran code are very limited and correspond to calls to a few *ComLib* routines, so it will be easy for a MésoNH user to add his own source modifications without problems concerning the parallelization.

4 The grid nesting technique

4.1 Principles

MésoNH is able to perform several simultaneous simulations on a nested grid. This technique (called grid nesting[2]) allows focusing on specific regions described by a higher spatial resolution, preserving a correct representation of large scale flow with a moderate size memory occupation and greater computational efficiency (an example is shown in Fig. 3). More than two models can be used together. The nesting is restricted to horizontal directions. The temporal and horizontal spatial resolution ratios between models must be integer.

The interaction between the fine mesh (child) model and its coarse mesh (parent) model are treated through either one- or two-way interactive nesting:

- In the simpler one-way approach, only waves coming from the parent model are allowed to enter and affect the child model. It is performed by the use of interpolated coarse fields to produce boundary conditions for the child model.

- In the two-way approach, waves resolved by the child model can also affect the parent model solution. It is done by relaxing the parent fields towards the average of the child fields over the overlapping domain.

4.2 Interpolation and averaging operators

The horizontal interpolation formula I at a point of coordinate (a, b) on the child grid uses the 16 nearest parent points and can be written as:

$$I(a, b) = \sum_{i=1}^{4} \sum_{j=1}^{4} f_{ij} \, B_i(a) \, B_j(b) \tag{1}$$

where f_{ij} corresponds to the parent value at points (i, j) and the B_i are third order polynomials obeying the property that the interpolation surface passes exactly through the 4 innermost points.

170

The horizontal averaging procedure A is a simple arithmetical average including all the points of the child grid embedded in the coarse parent mesh; the number of child points depends on the x and y resolution ratios:

$$A(\psi) = \frac{1}{X_{\text{ratio}} \cdot Y_{\text{ratio}}} \sum_{i=1}^{X_{\text{ratio}}} \sum_{j=1}^{Y_{\text{ratio}}} \psi_{i,j} \qquad (2)$$

where $\psi_{i,j}$ is the field on the child grid.

4.3 Parallel implementation

To achieve the parallelization of a nested grid run, two problems had to be solved:

- managing the decomposition of several nested domains, and

- parallelizing the parent–child interaction routines involving A and I.

One solution could be to use a coherent partitioning between the different models, i.e., one processor manages the same geographical area for all nested models. This solution could allow a simple treatment of model interaction. The sequential code of operators A and I could work in the same manner in a parallel run; no interprocessor communication would be needed. This solution is not used because it leads to very poor load balancing among the processors: some processors would not treat a subdomain belonging to the child model and would be unused during an important part of the computation.

The chosen solution is to independently and equitably distribute each model on all the available processors. Each processor manages different subdomains without spatial connection. To assure a correct communication inside a nested model, the Fortran types used in *ComLib* are recursive, meaning they contain pointers to variables of the same type. A special subroutine is called to indicate when the computation (and the associated interprocessor exchanges) switches from a parent model to its child.

The parallelization of operators A and I requires additional interface routines to perform data exchanges between parent and child models, both decomposed over processors. In *ComLib*, the list of zones to send is found by performing the intersection between the parent subdomain managed by the processor and all subdomains of the child domain. These intersections may be empty. For the child point of view, all the processors receive in an intermediate array the data of the parent model required for the interpolation. This area corresponds to the whole child subdomain but with the coarse resolution of the parent model. The lists of zones are hidden in a *ComLib* data structure and the user has only to allocate the intermediate array and specify which fields have to be communicated between the parent and child models. A *ComLib* routine fills this area with the values of corresponding parent data by communication between processors. Then, this intermediate array can be used in Eq. (1) as f_{ij}.

To parallelize the averaging operator, an inverse way is used. For each processor, Eq. (2) is applied to the child fields and the result is stored in a same-shape intermediate array. Another *ComLib* routine sends this array to a second intermediate array corresponding to the intersection of the parent subdomain and the child domain. Some processors receive no data because this intersection is empty. To use this second array for the two-way interaction, a *ComLib* routine returns the coordinates of the intersection converted to local coordinates. Averaging child fields before processor communication allows minimizing the amount of data exchange.

5   Results and performance

Figure 3 shows an example of a grid nesting simulation. The meteorological situation corresponds to the first severe storm which affected France at the end of 1999. The parent model resolution is 40 km and the grid is composed by 180×145×45 points. The child model has a fourfold finer resolution (10 km). The simulation was made on the Fujitsu VPP5000 at Météo-France. It

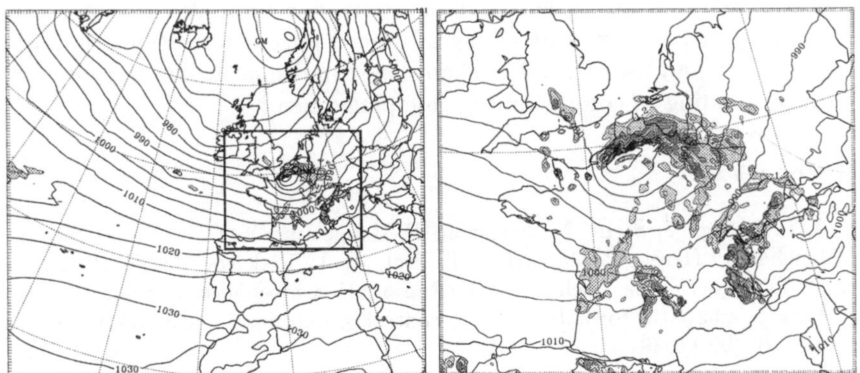

Figure 3. Rainfall and surface pressure (hPa) for the parent (left) and child (right) models at 06UTC 26 december 1999. The minimum value for the child pressure field is 966 hPa.

validates the parallel implementation of the grid nesting technique. Only small discrepancies between the one- and four-processor runs were found, which are attributable to compiler optimizations. The use of a stripe decomposition instead a rectangular one provides better efficiency, because the VPP5000 has vector processors, and the vectorization of the computation is essentially done on the first dimension of Fortran arrays. The scalability (i.e., the efficiency of parallel computing) of the pressure solver part is not very good at the present time because of the nonlocal method used to solve the elliptic equation. Without taking into account the time spent in the solver part, the efficiency of the parallel code is good (about 0.8 for an 8-processor run); a part of this number can be explained by a decrease in vectorization connected to a decrease in the size of the Fortran arrays. The cost of interaction between child and parent models increases with the number of processors but remains small in relation to the total model computation time.

6 Conclusion

The optimization of the parallel MésoNH version is still in progress. Most of the time, the interpolation procedure I is not used on the whole child subdomain but only on its physical boundaries. New *ComLib* routines have been developed which only communicate the necessary data to reduce the time spent in the communication. The technique used to solve the pressure equation will perhaps be changed to attain better parallel efficiency. Nevertheless, the primary goal of the work of parallelization is reached: to provide a research tool not only able to produce very fine simulations (using a huge number of grid points and including advanced parameterizations) but which is also flexible and user friendly. These two conditions are important to help the understanding of complex mesoscale phenomena.

References

1. J. P. Lafore, J. Stein, N. Asencion, P. Bougeault, V. Ducrocq, J. Duron, C. Fischer, P. Hereil, P. Mascart, J. P. Pinty, J. L. Redelsperger, E. Richard and J. Vila-Guerau de Arellano: The Méso-NH atmospheric simulation system. Part I: adiabatic formulation and control simulations. *Annales Geophysicae* **16**, 90–109 (1988)
2. T. L. Clark, Severe downslope windstorm calculations in two and three spatial dimensions using anelastic interactive grid nesting: a possible mechanism for gustiness. *J. Atmos. Sci.* **41**, 329–350 (1984)

NATURAL CONVECTION IN A CUBICAL POROUS CAVITY: SOLUTION BY ORTHOGONAL COLLOCATION

HUGO JIMÉNEZ-ISLAS

Departamento de Ingeniería Bioquímica, Instituto Tecnológico de Celaya,
38010 Celaya, Gto., México
E-mail: hugo@itc.mx

A numerical study has been performed to solve the problem of 3D natural convection in a cubical cavity containing an isotropic saturated porous medium with internal heat generation. The governing equations were transformed to a vector potential formulation and the resulting model was discretized using orthogonal collocation with Legendre polynomials. The resulting sets of nonlinear algebraic equations were solved by nonlinear relaxation. This study considers the influence of Rayleigh number and internal heat generation on isotherms and Nusselt number. The accuracy of the numerical method was verified via an energy balance with maximum errors in the order of 2 %.

1   Introduction

Natural convection in closed cavities containing isotropic saturated porous media has been of great interest in recent years, due to the numerous applications such as grain storage in silos, drying processes, exothermic reactions in packed beds, geothermal studies, heat transfer in nuclear waste storage, etc. Due to these, various studies have been published, analyzing the effects of the buoyancy forces and heat transfer in porous media with internal heat generation, most of them dealing with 2D geometries. Prasad[1] investigated natural convection in a 2D rectangular cavity with symmetric boundary conditions, finding that Nusselt number increases as Rayleigh number increases, or decreases as the geometrical aspect ratio increases. Prasad and Chui[2] studied 2D natural convection in a cylinder, analyzing the effects of diverse boundary conditions, finding a similar behavior for Nusselt number as Prasad[1]. Singh and Thorpe[3] studied natural convection and mass transport in grain storage in silos. These authors propose a mathematical model using the Volumetric Averaging Theorem[4,5], which can be applied to systems with arbitrary geometries. Pallares *et al.*[6] have employed the control volume scheme for study the flow structures in a cubical cavity filled with air and heated from below, for analyzing the Rayleigh–Bènard convective regime.

In this context, the purpose of this work is the numerical study of the natural convection in a cubical cavity containing a porous medium with internal heat generation. In this problem, boundary conditions were selected such that the phenomenon is three dimensional, analyzing the behavior of temperature patterns and the global heat transfer in the cavity, focusing the problem on the elucidation of transport processes in grain storage in silos.

Also, in the numerical solution of natural convection in a porous medium, we assessed the method of Orthogonal Collocation, implemented by Villadsen and Stewart[7], for solving differential equations with boundary values problems like viscous flow and heat and mass transfer with chemical reaction. This method has been fully described and applied to diverse problems in chemical engineering[7,8]. Actually, the problem of 3D natural convection has been solved numerically by well-established methods as finite differences, which requires a high number of nodes to obtain solutions with a good accuracy, at high computational cost.

2   Mathematical model

The system under study is a cubical closed cavity of dimension L shown in Fig. 1, containing an isotropic porous medium, with internal heat generation at a constant rate, Q_o. The bottom of the cavity is insulated, and the walls located at $x = 0$ and $z = 0$ are at temperature T_h; the top wall and the walls located at $x = L$ and $z = L$ are at temperature T_c, being T_h hotter than T_c. The internal heat source together with the temperature gradients at the boundaries will induce the movement by natural convection of the fluid through the porous medium. The choice of boundary conditions is a previously used approximation of the behavior of a cubical silo containing a cereal grain[9].

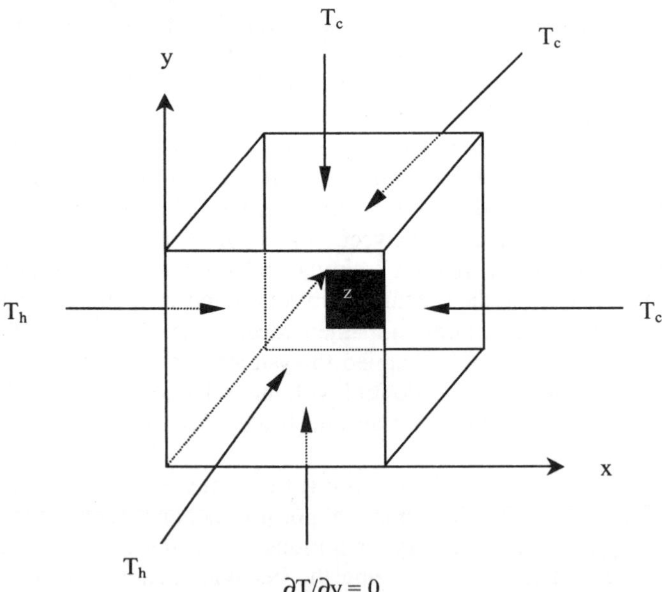

Figure 1 Geometrical system used in this study with boundary conditions

The equations that govern the transport of momentum and heat for a two-phase medium are[4,5,9]:

Continuity:

$$\nabla \cdot \mathbf{v} = 0 \qquad (1)$$

Darcy Law:

$$\mathbf{v} = -\frac{\mathbf{C}}{\mu} \cdot [\nabla \mathbf{p} - \rho \mathbf{g}] \qquad (2)$$

Energy:

$$(\rho C_v)(\mathbf{v} \cdot \nabla T) = \mathbf{K} : \nabla \nabla T + Q_0 \qquad (3)$$

To develop the final model, the following considerations have been made:

1. An effective, isotropic, darcian medium[10], with constant thermodynamic properties. In this case, tensor $\mathbf{C}$ of Darcy's law is transformed into scalar K, which is the classical term representing the permeability. Similarly, for the energy equation the conductivity tensor $\mathbf{K}$ is transformed into the effective thermal conductivity k_{eff} of the porous medium.
2. The cavity has rigid and impermeable walls therefore; there is no fluid slip.
3. The interstitial fluid is Newtonian, flowing in a steady-state laminar flow.

Based on these assumptions, the governing equations are rendered dimensionless so that the characteristic parameters of the problem (Ra and S_o) may be independently varied. In this work, we used the method of vector potential[11], which includes the exact satisfaction of the continuity equation and the elimination of the pressure term, because solving for this variable is not required. Hence, the mathematical model is:

Vector potential:

$$\mathbf{u} = \nabla \times \varphi \qquad (4)$$

Momentum:

$$\nabla^2 \varphi = -\text{Ra} \, (\nabla \times \theta_j) \qquad (5)$$

Energy:

$$(\nabla \times \varphi) \cdot \nabla \theta = \nabla^2 \theta + S_o \qquad (6)$$

Where $X = x/L$, $Y = y/L$, $Z = z/L$, $\mathbf{u} = \mathbf{v}L/\alpha$, $\theta = (T-T_c)/(T_h-T_c)$, $\text{Ra} = g\beta K(T_h-T_c)\alpha v$ and $S_o = Q_o L^2/ k_{eff} (T_h-T_c)$. To establish the boundary conditions for the vector potential v, we employed the impermeability state, such as described by Hirasaki and Hellums[12], while the boundary conditions for dimensionless temperature were:

$$\theta = 1 \quad \text{at X=0;} \qquad \theta = 0 \ \text{at X=1} \qquad (7a,b)$$
$$\partial\theta/\partial Y = 0 \ \text{at Y=0;} \qquad \theta = 0 \ \text{at Y=1} \qquad (7c,d)$$
$$\theta = 1 \quad \text{at Z=0;} \qquad \theta = 0 \ \text{at Z=1} \qquad (7e,f)$$

3   Orthogonal collocation in three dimensions

Equations (5) and (6) were discretized using the method of orthogonal collocation in three dimensions with Legendre polynomials. The application of this method has been described elsewhere[7,8,10] for one and two dimensions, which have been extended here to consider three dimensions, using matrices A and B accordingly. Thus, the first and the second derivatives are approximated as:

$$\frac{\partial \Theta}{\partial X} = \sum_{p=1}^{NX+2} AX_{ip}\,\Theta_{pjk} \qquad\qquad \frac{\partial^2 \Theta}{\partial X^2} = \sum_{p=1}^{NX+2} BX_{ip}\,\Theta_{pjk} \qquad\qquad (8\text{a,b})$$

$$\frac{\partial \Theta}{\partial Y} = \sum_{p=1}^{NY+2} AY_{jp}\,\Theta_{ipk} \qquad\qquad \frac{\partial^2 \Theta}{\partial Y^2} = \sum_{p=1}^{NY+2} BY_{jp}\,\Theta_{ipk} \qquad\qquad (8\text{c,d})$$

$$\frac{\partial \Theta}{\partial Z} = \sum_{p=1}^{NZ+2} AZ_{kp}\,\Theta_{ijp} \qquad\qquad \frac{\partial^2 \Theta}{\partial Z^2} = \sum_{p=1}^{NZ+2} BZ_{kp}\,\Theta_{ijp} \qquad\qquad (8\text{e,f})$$

In expressions (8a-f), NX, NY and NZ are the interior collocation points and Θ is the dependent variable. After discretization of equations (5) and (6), the resulting set of nonlinear algebraic equations was simultaneously solved by a nonlinear relaxation method[9] using an initial vector: $X° = [0, 0, 0, 0]^t$, with a convergence criterion of 10^{-5} for each equation. Later, Eq. (4) may be used for compute the velocity field. The numerical simulations were performed using the code ELI-COL3, developed by Jiménez-Islas[9] to solve nonlinear elliptic PDEs. The computational study considers a range of values of 10 to 1000 for Rayleigh numbers for a porous medium and 0 to 1000 for the magnitude of the dimensionless heat source. The values of the parameters and boundary conditions were selected because of their applicability to the storage of cereal grains in silos[4]. The simulations were carried out on an SGI Origin 2000[TM] supercomputer with MIPS R10000 processors running IRIX[TM].

4   Results and discussion

Table 1 shows some solved cases in this study. The interior collocation points were selected, ranging from 11×11×11 for low Rayleigh and S_o values, to 19×19×19 for higher Ra and S_o values; within the range of the expected behavior, since increasing Ra or S_o increases buoyancy forces and the distortion of the isotherms at the boundary increases, which may cause numerical instability[9]. As expected, temperature profiles are affected by the increase in the magnitude of the buoyancy forces (given in terms of the Rayleigh number Ra and the dimensionless heat generation S_o). At low Ra values (<10), the isotherms present little distortion, while conductive heat transport is dominant. For larger values of Ra (>10), the convective effects become important, causing the isotherms to be increasingly distorted.

Table 1. Some cases solved in this work with CPU time

Ra	S_o	OC Internal Points	w^a	Iterations	CPU time (s)
10	0	11×11×11	0.5	3,161	462
10	100	11×11×11	0.5	3,136	458
10	500	11×11×11	0.5	2,732	400
10	1000	15×15×15	0.5	5,035	2,947
100	0	11×11×11	0.5	3,253	490
100	100	11×11×11	0.5	2,993	453
500	0	15×15×15	0.3	9,755	5,717
100	100	15×15×15	0.1	4,385	2,573

^{a} Relaxation factor

Figure 2 shows the effect of the heat generation rate S_o on the isotherms, at a constant Ra = 10. When S_o = 0, the isotherms show only the effect of the buoyancy forces originated by the difference in temperatures at the vertical walls of the cavity. When S_o > 0, a hot nucleus starts to form and moves towards the upper part of the cavity, as the value of S_o is increased. This displacement is produced by the combined effect of S_o and Ra. A similar behavior is observed for Ra = 100, although in this case the hot kernel is nearer to the wall at X = 0. In order to appreciate more clearly the three-dimensional behavior of the isotherms, the results of the run for Ra = 100 and S_o = 100 are shown in Fig. 3. In this figure the presence of a hot core in the upper part of the cavity, originated by the volumetric heat source, can be observed. This core is surrounded by a cooler region ($\theta \cong 2.5$) that extends down to the lower part of the cavity. If the porous material were cereal grain stored in a silo, grain deterioration would be expected in this zone.

5 Global heat transfer

To define the Nusselt number, the difference between the average temperature θ_{med} in the cavity and the wall temperature θ_c is used. Therefore, starting from a global balance of energy, the average Nusselt number is defined as:

$$Nu_{av} = \frac{S_o - \int_0^1\int_0^1 \frac{\partial\theta}{\partial X}\Big|_{X=0} dYdZ - \int_0^1\int_0^1 \frac{\partial\theta}{\partial Z}\Big|_{Z=0} dXdY}{3\theta_{med}} \tag{9}$$

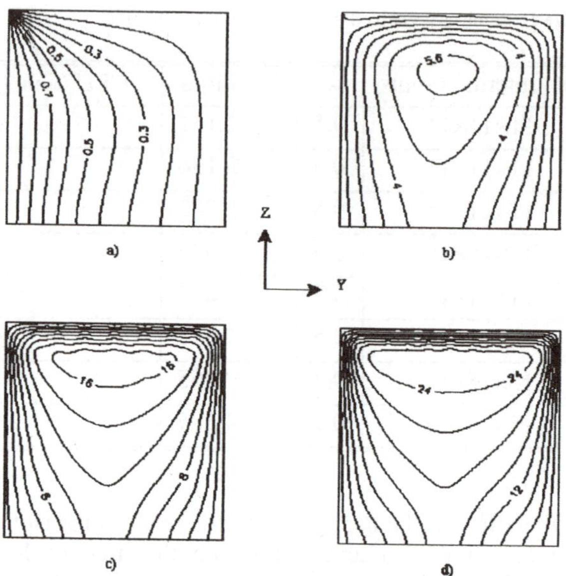

Figure 2. Isotherms at plane X = 0.5, for Ra = 10. (a) S_o = 0, (b) S_o = 100, (c) S_o = 500, (d) S_o = 1000

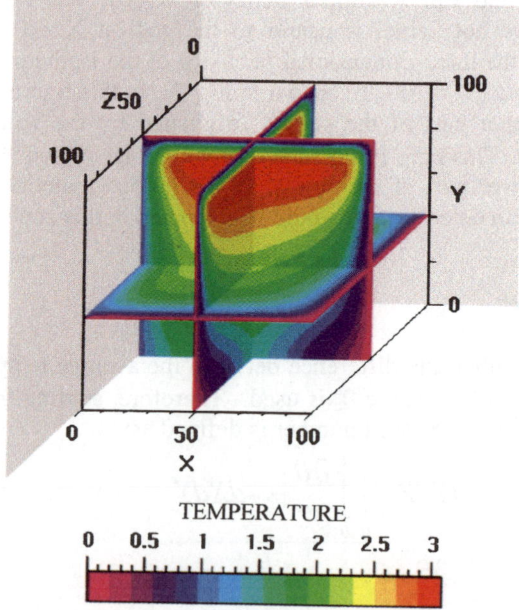

Figure 3. Intersection of XY, XZ and YZ planes at half of cavity, for Ra = 100 and S_o = 10

The values of θ_{max}, θ_{med} and Nu_{av} for the simulations reported in Table 1 are presented in Table 2, together with the percentage error in the global energy balance closure. The values obtained confirm that the number of interior collocation points is adequate, making this method suitable for solving this type of systems of elliptic PDEs. A comparison of this method versus finite differences for a 2D case is reported by Jiménez-Islas *et al.* [10]

Table 2 Average Nusselt Number and % error in energy balance

Ra	S_o	θ_{max}	θ_{med}	Nu_{av}	% Error
10	0	1.00	0.40	9.27	0.07
10	100	5.79	2.66	9.56	2.38
10	500	16.18	8.05	14.66	2.26
10	1000	26.75	12.97	17.85	1.40
100	0	1.00	0.37	12.26	0.12
100	100	2.98	1.65	17.31	1.16
500	0	1.00	0.37	22.42	1.14
100	100	2.02	1.16	22.08	0.35

It can be observed that Nu_{av} increases when values of Ra or S_o increase. This effect has been discussed in previous sections. Prasad[1] and Prasad and Chui[2] have reported similar behavior for 2D geometries.

6   Conclusions

A nonlinear elliptic PDE model describing natural convection in a cavity containing a porous medium has been solved numerically by the method of orthogonal collocation in three dimensions, using a relatively small number of interior collocation points (11 to 19). The code ELI-COL3 is not restricted to the number, type of boundary conditions or nonlinearity of the set of elliptic PDEs. In the numerical study, it is observed that there is a combination of buoyant forces (Ra and S_o) that gives rise to a hot core, whose temperature diminishes when Ra increases.

With respect to orthogonal collocation, we developed the algorithms to approximate the derivatives in 3D problems. We also determined that orthogonal collocation is a powerful method for solving sets of nonlinear PDEs using fewer nodes than those required by finite differences to obtain similar accuracy. Current results indicate that the method is competitive with other well-established algorithms. The solution of this problem gives the basis for the modeling of important technological applications, such as dynamics of grain storage in silos, storage of agricultural products and thermodynamical design of chemical packed-bed reactors.

7   Acknowledgements

The author is grateful to DGSCA–UNAM for access to the SGI Origin 2000 supercomputer and to Dr. Felipe López-Isunza for early comments on this work.

References

1. V. Prasad, "Thermal convection in a rectangular cavity filled with a heat-generating, Darcy porous medium", *J. of Heat Transfer* **109**, 697–703 (1987).
2. V. Prasad and A. Chui, "Natural convection in a cylindrical porous enclosure with internal heat generation", *J. of Heat Transfer* **111**, 916–925 (1989).
3. K. Singh and G. R. Thorpe, "A solution procedure for Three-dimensional free convective flow in peaked bulks of grain", *J. Stored Prod. Res.* **29**(3) 221–235 (1993).
4. R. G. Carbonell and S. Whitaker, "Heat and mass transport in porous media", in *Fundamentals of transport phenomena in porous media*, J. Bear and M.Y. Corapcioglu, eds., 121–198 (Martinus Nijhoff Publishers, Dordrecht, 1984).
5. S. Whitaker, "Flow in porous media I: A theoretical derivation of Darcy's Law", *Transport in Porous Media* **1**, 3–25 (1986).
6. J. Pallares, I. Cuesta, F. X. Grau and F. Giralt, "Natural convection in a cubical cavity heated from below at low Rayleigh numbers", *Int. J. Heat Mass Transfer* **39**, 3233–3247 (1996).
7. J. V. Villadsen and W. E. Stewart, "Solution of boundary-value problems by orthogonal collocation", *Chem. Engng. Sci.* **22**, 1483–1501 (1967).
8. B. A. Finlayson, *Nonlinear analysis in chemical engineering*, chap. 4–6 (McGraw-Hill Book Co. , New York, 1980).
9. H. Jiménez-Islas, *Modelamiento de los procesos de transferencia de momentum, calor y masa en medios porosos*, Ph.D. Thesis (Universidad Autónoma Metropolitana—Iztapalapa, México, 1999).
10. H. Jiménez-Islas, F. López-Isunza and J. A. Ochoa-Tapia, "Natural convection in a cylindrical porous cavity with internal heat source: a numerical study with Brinkman-extended Darcy model", *Int. J. Heat Mass Transfer* **42**, 4185–4195 (1999).
11. A. K. Singh, E. Leonardi and G. R. Thorpe, "Three-dimensional natural convection in a confined fluid overlying a porous layer", *J. of Heat Transfer* **115**, 631–638 (1993) .
12. G. J. Hirasaki and G. D. Hellums, "A General formulation of the boundary conditions on the vector potential in three-dimensional hydrodynamics", *Quarterly of Applied Mathematics* **XXVI**, 331–342 (1968).

NUMERICAL SOLUTIONS OF NONLINEAR DIA EQUATIONS AND CALCULATION OF TURBULENT DIFFUSIVITIES

N. A. SILANT'EV[*]

*Instituto Nacional de Astrofísica, Óptica y Electrónica,
Apartado Postal 51 y 216, 72000 Puebla, Pue., México
E-mail: silant@inaoep.mx*

The exact numerical solutions of nonlinear direct interaction approximation equation or set of equations (DIA equations for Green's functions), describing the propagation of scalar impurities and magnetic field in a turbulent medium, are presented. It is assumed that the turbulence is stationary, isotropic and homogeneous. The steady state values of turbulent diffusivities and alpha coefficient are calculated. It is shown that such approach allows us to obtain the satisfactory values of turbulent transport coefficients for all the values of turbulent Strouhal numbers.

1 Introduction

The problem of turbulent diffusion of passive fields (number density, temperature and magnetic field) is of great importance in astrophysics, geophysics, meteorology and hydrodynamics. Usually one is interested only in the knowledge of turbulent diffusivities D_T if Euler's picture of turbulent motions is known. The molecular diffusivity D_m is much smaller than the value D_T and can be neglected when finding D_T. The value D_T depends on the behavior of the turbulence as a whole (the form of the spectrum, particular dependence of the velocity correlations on the time, etc.). But mostly the value D_T depends on the parameters u_0, τ_0 and R_0 which are the characteristic velocity, lifetime and correlation space scale of turbulent motions, respectively.

The qualitative estimates of D_T can be derived from the well known diffusion relation $L^2 \simeq D_T \, t$ where instead of the time t one can take the lesser of two characteristic times τ_0 and $t_0 \simeq R_0/u_0$ (t_0 is called the "turnover" time). In this case the value L corresponds to the free length of the liquid particle in turbulent motion. If $\tau_0 \ll t_0$ (or the turbulent Strouhal number $\xi_0 \equiv u_0\tau_0/R_0 \ll 1$), one has $L \simeq u_0\tau_0$ and $D_T \simeq u_0^2\tau_0$. In the opposite limit $\tau_0 \gg t_0$ (or $\xi_0 \gg 1$) one has $L \simeq u_0 t_0 \simeq R_0$ and the value $D_T \simeq u_0 R_0$ is independent of τ_0.

But how to choose the parameters τ_0 and R_0 from the known form of the turbulent spectrum? This problem is clearly resolved if the spectrum is of peak-like type. If the spectrum has a broad form, or has several peaks, such

---

[*]Permanent address: Pulkovo Astronomical Observatory, St. Petersburg 196140, Russia

a choice is rather difficult. The theory of turbulent diffusion is to calculate the transport coefficients, like D_T, without using such notions as τ_0 and R_0, directly from turbulent spectra.

2 Basic equations

Linear stochastic equations for the number density $N(\mathbf{r}, t)$ of some impurity and the magnetic field $\mathbf{B}(\mathbf{r}, t)$ have the form:

$$\left(\frac{\partial}{\partial t} - D_m \nabla^2\right) N(\mathbf{r}, t) = -\nabla[\mathbf{u}(\mathbf{r}, t) N(\mathbf{r}, t)] \equiv LN(\mathbf{r}, t), \tag{1}$$

$$\left(\frac{\partial}{\partial t} - D_m \nabla^2\right) \mathbf{B}(\mathbf{r}, t) = \nabla \times (\mathbf{u}(\mathbf{r}, t) \times \mathbf{B}(\mathbf{r}, t)) \equiv L\mathbf{B}(\mathbf{r}, t). \tag{2}$$

Here, $\mathbf{u}(\mathbf{r},t)$ is the turbulent Euler velocity of the medium. Below we use the convenient notations: $f(n) \equiv f(\mathbf{r}_n, t_n)$, $dn \equiv d\mathbf{r}_n dt_n$, $\mathbf{R} = \mathbf{r}_1 - \mathbf{r}_2$, $\tau = t_1 - t_2$, etc. For brevity we use the operator notation L as defined in Eqs. (1) and (2). The ensemble of turbulent velocity $\mathbf{u}(\mathbf{r}, t)$ is assumed to be homogeneous, isotropic and stationary and the mean value $\langle \mathbf{u}(\mathbf{r}, t)\rangle = 0$. All the values considered are represented as a sum of mean value and fluctuating part, e.g., $\mathbf{B} = \mathbf{B}_0 + \mathbf{b}$ with $\langle \mathbf{B}\rangle = B_0$ and $\langle \mathbf{b}\rangle = 0$.

The averaging of initial equations gives rise to the coupled system of equations for the mean and fluctuating parts of number density or magnetic field. This means that if we want to write the separate equation only for mean value, then such an equation can be represented as an infinite hierarchy of nonlinear equations with growing degree of nonlinearity. In reality such hierarchy can be written only for the averaged part of the Green function of Eqs. (1) or (2), not for the values $N_0(\mathbf{r}, t)$ and $\mathbf{B}_0(\mathbf{r}, t)$. The simplest nonlinear equation, with the second order nonlinearity, was proposed and studied by Kraichnan[1] and Roberts[2]. It was called the direct interaction approximation equation or DIA equation for short. The simple derivation of the whole hierarchy of nonlinear equations both for scalar and vector passive fields was given by Dolginov and Silant'ev[3] and Silant'ev[4]. The first equation of this hierarchy coincides with the DIA equation. We present here the first three terms of this hierarchy for the case of the scalar Eq. (1):

$$\left(\frac{\partial}{\partial t_1} - D_m \nabla_1^2\right) G(1 - 2) = \delta(\mathbf{R})\delta(\tau) + \int d3 \langle L(1)G(1 - 3)L(3)\rangle G(3 - 2)$$

$$+ \int d3 \int d4 \langle L(1)G(1 - 3)L(3)G(3 - 4)L(4)\rangle G(4 - 2)+$$

$$\int d3 \int d4 \int d5 \left\{ \langle L(1)G(1-3)L(3)G(3-4)L(4)G(4-5)L(5)\rangle\, G(5-2) - \right.$$

$$\langle L(1)G(1-3)\,\langle L(3)G(3-4)L(4)\rangle\, G(4-5)L(5)\rangle\, G(5-2) -$$

$$\left. \langle L(1)G(1-3)L(3)\rangle\, G(3-4)\,\langle L(4)G(4-5)L(5)\rangle\, G(5-2) \right\} + \cdots \tag{3}$$

Here $\langle G(1;2)\rangle \equiv G(1-2)$ is the averaged Green's function of Eq. (1). For Gaussian ensemble of $\mathbf{u}(\mathbf{r},t)$ the second integral term vanishes and the third reduces to the single term, where $L(1)$ is averaged with $L(4)$ and $L(3)$ is averaged with $L(5)$. Dropping in Eq. (3) all the terms of the hierarchy except the first integral, we obtain the DIA equation. For the magnetic field case the hierarchy looks like Eq. (3) but the Green's function and the operator of interaction L are now tensors. The sequence of the terms of the hierarchy (3) describes the turbulent transport of passive fields with increasing degree of detail.

From Eq. (3) follows the formally exact integrodifferential equation for the mean number density $N_0(\mathbf{r},t)$. Assuming that $N_0(\mathbf{r},t)$ is a smooth function on the scales $\sim R_0$ and the times $\sim \tau_0$ (or t_0), one can take in integral term $N_0(\mathbf{r}-\mathbf{R}, t-\tau)\simeq N_0(\mathbf{r},t)-(\mathbf{R}\nabla)N_0(\mathbf{r},t)$ and obtain the approximate diffusion equation for $N_0(\mathbf{r},t)$ with the renormalized diffusivity:

$$\left(\frac{\partial}{\partial t} - (D_m + D_T)\nabla^2\right) N_0(\mathbf{r},t) = 0. \tag{4}$$

For locally homogeneous, isotropic and stationary turbulence all the values parametrically depend on the position $\mathbf{r}$ and the time t.

The analogous renormalized diffusion equation for the mean magnetic field $\mathbf{B}_0(\mathbf{r},t)$ has the form:

$$\left(\frac{\partial}{\partial t} - D_m\nabla^2\right) \mathbf{B}_0(\mathbf{r},t) = \nabla\times\alpha(\mathbf{r},t)\mathbf{B}_0(\mathbf{r},t) - \nabla\times D_T(\mathbf{r},t)(\nabla\times\mathbf{B}_0(\mathbf{r},t)). \tag{5}$$

Here, the α coefficient describes the enhancement of the mean magnetic field in helical turbulence.

It should be stressed that the diffusion approximation does not need the τ expansion in $N_0(\mathbf{r}-\mathbf{R}, t-\tau)$ (see Silant'ev[5]), in contrast to the work of Moffatt[6] and Kraichnan[7].

Using the DIA equation one can obtain the following formulae for the steady-state turbulent diffusivity D_T of a scalar impurity:

$$D_T = \frac{1}{3}\int_0^\infty dp \int_0^\infty d\tau\{[\,E_{inc}(p,\tau) + E_{comp}(p,\tau)]\,g(p,\tau) +$$

$$E_{comp}(p,\tau)\, p\frac{\partial}{\partial p}\, g(p,\tau)\}. \qquad (6)$$

Here, $g(p,\tau)$ is the Fourier transform of the mean Green's function $g(\mathbf{R},\tau)$, and $E_{inc}(p,\tau)$ and $E_{comp}(p,\tau)$ are the spectra of incompressible and compressible turbulent motions, respectively (see Silant'ev[8] for more details):

$$\langle \mathbf{u}(\mathbf{r},t)\cdot\mathbf{u}(\mathbf{r},t+\tau)\rangle = \int_0^\infty dp[\,E_{inc}(p,\tau)+E_{comp}(p,\tau)]. \qquad (7)$$

For particular numerical calculations of the turbulent diffusivities D_T we have assumed that all the spectra are proportional to $\exp(-\tau/\tau_0)$. In this case for the calculation of D_T we need the Laplace transform of $g(p,\tau)$. Let us denote it as $g(p,s)$ with s being the Laplace transform variable. As a result the DIA equation for $g(p,s)$ acquires the form of the integral equation

$$g(p,s) = \Big\{ s + D_m p^2 + \frac{1}{4}\int_0^\infty dq \int_0^\infty d\tau \int_{-1}^{1} d\mu\times$$
$$\big[(1-\mu^2)p^2 E_{inc}(q,\tau) + 2\mu p(\mu p - q)\, E_{comp}(q,\tau)\big]\times$$
$$g(|\,\mathbf{p}-\mathbf{q}\,|,\tau)\exp(-s\tau)\Big\}^{-1}, \qquad (8)$$

where $\mu = \mathbf{p}\cdot\mathbf{q}/pq$ is the cosine of the angle between the vectors $\mathbf{p}$ and $\mathbf{q}$. The sequence of iterations of Eq. (8) constitutes a rapidly converging continued fraction. For the case of incompressible turbulence all the terms of this fraction are positive, which means that consecutive iterations represent the solution with alternating deficiency and abundance. It can be shown that the asymptotic solution of Eq. (8) for large p

$$g^{(0)}(p,s) = 2\{s + D_m p^2 + [(s + D_m p^2)^2 + 4/3 u_0^2 p^2]^{1/2}\}^{-1} \qquad (9)$$

represents the true value $g(p,s)$ with the deficiency. This value was used as an initial iteration. For compressible turbulence this technique also gives very rapid convergence.

For the magnetic field case the tensor Green function has the form:

$$G_{nm}(p,s) = \delta_{nm}G_0(p,s) + ie_{nmk}p_k G_1(p,s). \qquad (10)$$

The $G_1(p,s)$ function is equal to zero for the turbulence without helicity. The spectrum of helicity is determined by the relation:

$$\langle \mathbf{u}(\mathbf{r},t)\cdot\nabla\times\mathbf{u}(\mathbf{r},t+\tau)\rangle = \int_0^\infty dp\, E_h(p,\tau). \qquad (11)$$

The magnetic turbulent diffusivity D_T and α coefficient are described by the expressions:

$$D_T = \frac{1}{3} \int_0^\infty dp \int_0^\infty d\tau [\, E_{inc}(p,\tau)G_0(p,\tau) + E_h(p,\tau)G_1(p,\tau)], \qquad (12)$$

$$\alpha = \frac{1}{3} \int_0^\infty dp \int_0^\infty d\tau [p^2 E_{inc}(p,\tau)G_1(p,\tau) + E_h(p,\tau)G_0(p,\tau)]. \qquad (13)$$

If the turbulence does not possess any helicity, then $G_1 = 0$ and the DIA equation for G_0 coincides with Eq. (8) for scalar impurity. Of course, for this case one has $\alpha \equiv 0$.

The DIA system of nonlinear equations for $G_0(p,s)$ and $G_1(p,s)$ has a simpler form if we introduce the auxiliary functions $A(p,s) = G_0(p,s) + pG_1(p,s)$ and $B(p,s) = G_0(p,s) - pG_1(p,s)$:

$$A(p,s) = \left\{ s + D_m p^2 + \frac{p}{4} \int_0^\infty dq \int_0^\infty d\tau \int_{-1}^1 d\mu(1 - \mu^2)\exp(-s\tau)\times \right.$$

$$\left[pE_{inc}(q,\tau)G_0(\mid \mathbf{p} - \mathbf{q} \mid,\tau) - pE_h(q,\tau)G_1(\mid \mathbf{p} - \mathbf{q} \mid,\tau)+ \right.$$

$$\left. \left. (p^2 + q^2 - pq\mu)E_{inc}(q,\tau)G_1(\mid \mathbf{p} - \mathbf{q} \mid,\tau) - E_h(q,\tau)G_0(\mid \mathbf{p} - \mathbf{q} \mid,\tau)\right]\right\}^{-1}, \qquad (14)$$

$$B(p,s) = \left\{ s + D_m p^2 + \frac{p}{4} \int_0^\infty dq \int_0^\infty d\tau \int_{-1}^1 d\mu(1 - \mu^2)\exp(-s\tau)\times \right.$$

$$\left[pE_{inc}(q,\tau)G_0(\mid \mathbf{p} - \mathbf{q} \mid,\tau) - pE_h(q,\tau)G_1(\mid \mathbf{p} - \mathbf{q} \mid,\tau)- \right.$$

$$\left. \left. (p^2 + q^2 - pq\mu)E_{inc}(q,\tau)G_1(\mid \mathbf{p} - \mathbf{q} \mid,\tau) + E_h(q,\tau)G_0(\mid \mathbf{p} - \mathbf{q} \mid,\tau)\right]\right\}^{-1}. \qquad (15)$$

Here we have assumed that the turbulence is incompressible. This system is solved by iteration, using the simple asymptotics for $A(p,s)$ and $B(p,s)$.

3 Results and discussion

First we discuss the results of DIA calculations of turbulent diffusivities of scalar impurities. The main result is that the DIA equation allows us to calculate the turbulent diffusivities for all values of the Strouhal number $\xi_0 = u_0\tau_0/R_0$. For $\xi_0 \ll 1$ one can take $g(p,\tau) \simeq 1$ and Eq. (6) gives rise to

$$D_T = \frac{1}{3} \int_0^\infty dp \int_0^\infty d\tau [\, E_{inc}(p,\tau) + E_{comp}(p,\tau)] \simeq u_0^2\tau_0/3 = (u_0/p_0)\xi_0/3.$$

$$(16)$$

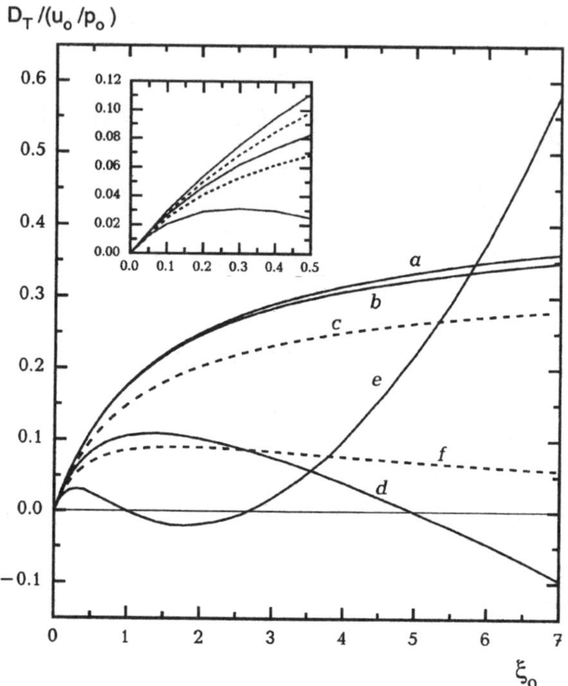

Figure 1. The dimensionless turbulent diffusivity $D_T/(u_0/p_0)$ for the case of the turbulence with the Kolmogorov type spectra $E_{inc,compr}(p,\tau) = (u_0^2/p_0)0.65159x^4(1 + x^{17/3})^{-1}\exp(-\tau/\tau_0)$ with $x = p/p_0$. The curves a, b and c represent, respectively, DIA values, DIA values corrected by the contribution of fourth-order velocity correlators, and self-consistent diffusivity, calculated from (6) with the diffusion Green's function $g(p,\tau) = \exp(-D_T p^2 \tau)$. They correspond to incompressible turbulence. The curves d, e and f are the corresponding diffusivities for the compressible (pure potential) turbulence with the same spectrum. The inset depicts the initial sections of the curves drawn to a larger scale.

Here we neglect the terms with small molecular diffusivity D_m. The numerical calculations show that this formula is valid up to $\xi_0 \approx 0.5$ and is independent of compressible or incompressible is the turbulence.

For $\xi_0 \gg 1$, which corresponds to "frozen" turbulence, one can assume $E_{inc}(p,\tau) \equiv E_{inc}(p)$. Using the asymptotic expression (9) we obtain the approximate formula:

$$D_T \simeq \frac{1}{\sqrt{3}} \int_0^\infty dp\, E_{inc}(p)/pu_0 \approx (u_0/p_0)/\sqrt{3}. \tag{17}$$

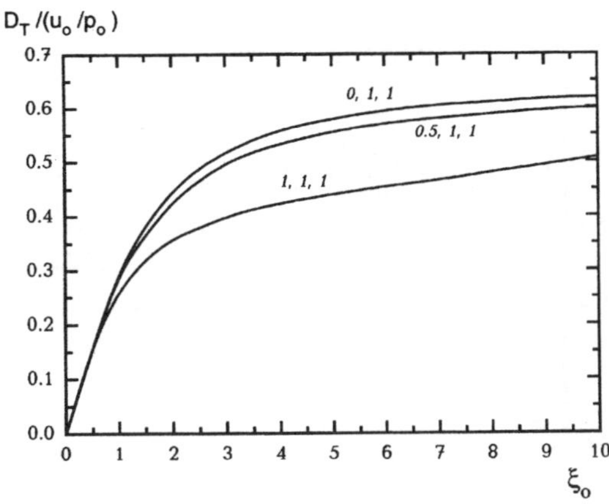

Figure 2. The dimensionless turbulent diffusivity of magnetic field $D_T/(u_0/p_0)$ for the case of incompressible turbulence with spectrum $E_{inc}(p,\tau) = u_0^2\,\delta(p - p_0)\exp(-\tau/\tau_0)$ and helicity spectrum $E_h(p,\tau) = H_0\,\delta(p - p_h)\exp(-\tau/\tau_h)$. The numbers denote the values of the parameters $a = H_0/u_0^2 p_0$ (the degree of helicity), τ_0/τ_h and p_h/p_0, respectively. The curves for the cases $(0,5,1)$, $(0,1,5)$, $(0,5,5)$, $(1,1,5)$, $(1,5,1)$ and $(1,5,5)$ are essentially the same as the represented case $(0,1,1)$.

In practice, the estimation (17) is valid for $\xi_0 > 5\text{–}10$. For the case of pure compressible (potential) turbulence of nonacoustical type (see Fig. 1) the diffusivity D_T acquires negative values for $\xi_0 > 3$. It does not mean that actual values of D_T are negative, indeed, because such large values of ξ_0 for the potential turbulence are impossible in nature (see Silant'ev[8]).

Why does the DIA equation allow us to calculate D_T for all values of ξ_0? The first reason is that the nonlinear DIA equation takes into account all the possible contributions of two-point velocity correlations, i.e., all the degrees of this correlator including those which constitute parts of fourth-order and higher correlators. The second reason is that just such forms of correlators describe the structure of large-scale turbulent motions which are of greatest importance for turbulent diffusion.

But what about the exactness of these DIA calculations? Using the formula for the contribution of the remaining fourth-order velocity correlators, which follows from the next term of our hierarchy, we conclude that the corrections are negative and grow monotonically from 0% at $\xi_0 = 0$ up to 10–11% (for peak-like spectra) and $\sim 7\%$ (for broad spectra) at $\xi_0 \to \infty$. These calculations assume that the ensemble of the turbulent velocities is gaussian.

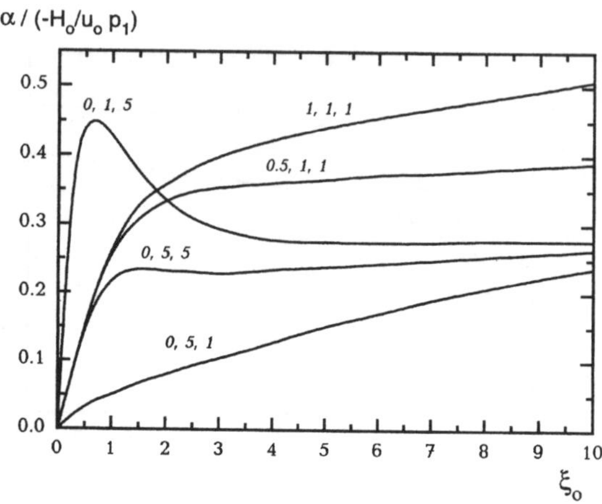

Figure 3. The dimensionless α coefficient $\alpha/(-H_0/u_0p_h)$ for the spectra of Fig. 2. The notations are the same as in Fig. 2. The case (0,1,5) essentially coincides with (1,1,5); the case (0,5,1) coincides with (1,5,1); the case (0,5,5) coincides with (1,5,5).

The DIA equation for scalar impurities does not depend on the helicity. Therefore we have used the next term of the hierarchy to estimate the dependence of D_T on the turbulent helicity. The helicity increases the D_T values (see Kraichnan[7], Silant'ev[9]). For a 100% helical medium (where $pE_{inc}(p,\tau) = E_h(p,\tau)$) the D_T coefficient can increase up to 50% (for $\xi_0 \to \infty$).

The set of equations (14) and (15) depends on the spectrum of the helicity. In this case the helicity decreases the turbulent diffusivity D_T of the magnetic field, in contrast to the case of scalar impurities. This is due to the fact that the helicity, because of the α-effect, increases the level of the mean magnetic field. This increase is larger than the corresponding increase in turbulent mixing, so the relative turbulent diffusion, described by D_T, is diminished by the helicity. The helicity decrease in D_T is very small for degrees of helicity $a = H_0/u_0^2 p_0 \leq 0.5$ (here $H_0 = \langle \mathbf{u}(\mathbf{r},t) \cdot \nabla \times \mathbf{u}(\mathbf{r},t) \rangle$ and $p_0 \approx 1/R_0$).

Some results of calculations of D_T and α are represented in Figs. 2 and 3. It is seen that the dimensionless value $\alpha/(-H_0/u_0p_h)$ increases monotonically with increasing ξ_0 for large scale helicity when the characteristic helicity wave number $p_h \approx p_0$. For $a \leq 0.5$ the dimensionless α practically does not depend on the helicity. For small-scale and (or) short-time correlated helicity ($p_h/p_0 \geq 5$ and (or) $\tau_0/\tau_h \geq 5$) the dimensionless α is essentially independent of the helicity. For small scale helicity with $p_0\sqrt{\tau_0/\tau_h}/p_h < 1$ the dimensionless α has a peak-like profile.

When the magnetic energy is comparable with the kinetic energy of the turbulence the back-reaction of the magnetic field changes the structure of the turbulent motions. In this case the spectra $E_{inc}(p,\tau)$ and $E_h(p,\tau)$ and the values u_0, R_0 and τ_0 parametrically depend on the mean magnetic field B_0. According to our general formulas (12) and (13) the values α and D_T also depend on B_0. For this case our calculations for the limiting cases of peak-like and broad spectra give the whole range of possible variations of α and D_T.

It should be stressed that only the nonlinear equations of hierarchy (3) allow the calculation of D_T and α for the entire range of the turbulent Strouhal number ξ_0. Numerous linear approaches did not succeed in this problem (see, for example, Carvalho[10]).

The solution of the problem for nonexponential time dependence of the turbulent spectra requires supercomputing calculations, especially if we want take into account the higher terms of hierarchy (3). At present, techniques for such calculations have not been developed.

References

1. R. H. Kraichnan, *J. Fluid Mech.* **5**, 497 (1959).
2. P. H. Roberts, *J. Fluid Mech.* **11**, 257 (1961).
3. A. Z. Dolginov and N. A. Silant'ev, *Geophys. Astrophys. Fluid Dyn.* **63**, 139 (1992).
4. N. A. Silant'ev, *Sov. Phys. JETP* **74**, 650 (1992).
5. N. A. Silant'ev, *Sov. Phys. JETP* **85**, 712 (1997).
6. H. K. Moffatt, *J. Fluid Mech.* **65**, 1 (1974).
7. R. H. Kraichnan, *J. Fluid Mech.* **75**, 657 (1976).
8. N. A. Silant'ev, *Sov. Phys. JETP* **87**, 505 (1998).
9. N. A. Silant'ev, *Geophys. Astrophys. Fluid Dyn.* **75**, 283 (1994).
10. J. C. Carvalho, *Astron. & Astrophys.* **261**, 348 (1992).

PART IV

APPLICATIONS

FROM STARS TO VOLCANOES: THE SPH STORY

J. J. MONAGHAN

Department of Mathematics and Statistics, Monash University,
Clayton, Vic. 3800, Australia
E-mail: joe.monaghan@sci.monash.edu.au

Smoothed Particle Hydrodynamics (SPH) was originally developed to handle multidimensional problems in astrophysics. Like all particle methods it makes use of the fact that advection follows the motion of the particles. As a consequence there are never problems with advection, and interfaces are automatically followed because particles define interfaces. In addition, the resolution of SPH varies naturally and automatically in space and time. These advantages carry over to a wide variety of fluid dynamical problems involving more than one phase and more than one material. In this paper I will describe how SPH can be extended to deal with problems involving solid bodies impacting fluids, debris flows hitting water, and a recent extension of Benz and Asphaug's[1] work on the fracture of brittle solids using SPH.

1 Introduction

SPH is a fully Lagrangian method that uses moving interpolation points which behave like particles (for early references see Monaghan[2], Benz[3] and for other meshless methods see Belytschko *et al.*[4]). Like all particle methods SPH exploits the fact that the characteristics of the advective (or hyperbolic) part of the equations of motion are the trajectories of the particles. Advection therefore never creates difficulties and this is one of the reasons that Harlow[5] devised the particle method PIC. Interfaces are automatically followed by particle methods because the particles define the interface. This does not, of course, mean that interfaces are always followed accurately, but the problem of defining properties at the interface is reduced to that of determining the interactions between the particles. Another nice feature of SPH is that the resolution automatically adjusts to the concentration of particles. For example, in a volcanic eruption the ash ends up in a dense layer on the ground while the hot gas rises into the atmosphere. In an SPH simulation of the eruption the ash and the hot gas can have different resolutions which vary in space and time. This is a real advantage, and incomparably simpler to implement than the nested grids used in some finite difference calculations.

Because particle methods are close to the underlying physics of the problem, they seldom give absurd answers. If you make a poor choice of parameters, particle methods will give the correct answer to a different physical problem. For example, in plasma physics simulations, a bad choice of parameters can cause the particle method to heat the gas rather than become unstable. In astrophysical simulations an initial feature (it might be an orbiting disk) might be initially too narrow for the resolution of the particle method. In that case SPH will cause the ring to spread radially (in a small fraction of the orbital time) until the resolution prescribed is adequate to describe the much slower viscous radial spreading (Murray et al.[6]).

In the case of shock dynamics of a perfect gas, Riemann and Total Variation Diminishing (TVD) methods can usually give more accurate solutions than SPH for a given amount of work. Nevertheless, even for these specialized problems, SPH gives good results which do not suffer from the errors that cause problems for some Riemann solvers (Monaghan[7]). Good results are also found for ultrarelativistic problems (Chow and Monaghan[8]). Where SPH has significant advantages is in more complex problems where it may be difficult or impossible to use methods based on Riemann solvers. These problems include the dynamics of fluids with free surfaces (Monaghan[9]), breaking waves (Monaghan and Kos[10,11]), the dynamics of systems containing several fluids (Monaghan et al.[12]), moving rigid bodies and elastic material which can fracture (Benz and Asphaug[1,13], Randles and Libersky[14], Monaghan[15]), and porous flow (Zhu et al.[16]).

1.1   Basic SPH interpolation

Interpolation from disordered particles is achieved by using the integral interpolant

$$A_I(\mathbf{r}) = \int A(\mathbf{r}')W(\mathbf{r} - \mathbf{r}', h)\mathbf{dr}', \tag{1}$$

where A can be a scalar, vector or general tensor field. The integration is over the available space, and the kernel W approximates a delta function. It is convenient to choose W from the class of functions which satisfy

$$\int W(\mathbf{r} - \mathbf{r}', h)\mathbf{dr}' = 1, \tag{2}$$

and

$$\lim_{h \to 0} W(\mathbf{q}, h) = \delta(\mathbf{q}). \tag{3}$$

Most current SPH calculations base W on the cubic or quartic spline. However, these are unlikely to be optimal and it would be worthwhile to search for better kernels. Different choices of kernel correspond to different choices of difference scheme or different finite element schemes. SPH therefore represents a class of methods.

The length h determines the resolution. In the initial set up it should be $\sim 1.5\Delta p$ where Δp is the particle spacing. During the motion it should be varied in a way which relates to the particle spacing. A simple way of achieving this is to take $h \propto 1/N^{1/d}$ where d is the number of dimensions and N is the particle number density. Usually N can be replaced by the mass density ρ. Steinmetz and Mueller[17] discuss alternative choices of the variation of h with density.

The integral interpolant can be approximated by a summation over mass elements. We then find the basic SPH interpolation formula[2]

$$A(\mathbf{r}) = \sum_b m_b \frac{A_b}{\rho_b} W(\mathbf{r} - \mathbf{r}_b, h), \tag{4}$$

where the summation is over those particles b which contribute to W. In practice W has compact support so that, with a good choice of h, a small number (typically about 40 in three dimensions) of particles contribute to the summations.

If W is a differentiable function then the interpolant may be differentiated analytically. In SPH the spatial gradients are obtained by the *exact* differentiation of an *approximate* interpolation formula.

1.2   The equations of gas dynamics

Using the above interpolation formula, the equations for the dynamics of a fluid with pressure P, mass density ρ, velocity $\mathbf{v}$, and thermal energy per unit mass u can be written as ordinary differential equations for particles[2]. For example the change of density for particle a can be written

$$\frac{d\rho_a}{dt} = \sum_b m_b \mathbf{v}_{ab} \cdot \nabla_a W_{ab}, \tag{5}$$

where $\mathbf{v}_{ab}$ denotes $(\mathbf{v}_a - \mathbf{v}_b)$, and W_{ab} denotes $W(\mathbf{r}_a - \mathbf{r}_b, h)$. Here, and elsewhere, ∇_a denotes the gradient taken with respect to the coordinates of particle a. One form of the acceleration equation for particle a is

$$\frac{d\mathbf{v}_a}{dt} = -\sum_b m_b \left(\frac{P_a}{\rho_a^2} + \frac{P_b}{\rho_b^2} + \Pi_{ab} \right) \nabla_a W_{ab}, \tag{6}$$

where the symmetry guarantees linear and angular momentum conservation. Another form of the acceleration equation which also guarantees exact momentum conservation is similar to the above but the pressure terms are replaced by

$$\frac{P_a + P_b}{\rho_a \rho_b}.$$

(7)

This latter form is more stable when two materials with greatly different density and pressure (for example a liquid and air) are in contact. Incidentally, this reminds us again that SPH represents a class of algorithms all of which use the basic interpolation, but the way in which the equations are written can vary and some forms are much better than others.

Π_{ab} provides the viscosity Monaghan[2,9], Monaghan and Gingold[18]. When real viscosities are required the form chosen by Cleary[19], based on the conduction term of Cleary and Monaghan[20], is reliable even when adjoining materials have viscosities which differ by several orders of magnitude.

The equation for the rate of change of energy per unit mass u can be written

$$\frac{du_a}{dt} = -\sum_b m_b \left(\frac{P_a}{\rho_a^2} + \frac{1}{2}\Pi_{ab} \right) \mathbf{v}_{ab} \cdot \nabla_a W_{ab},$$

(8)

and the particles are moved according to

$$\frac{d\mathbf{r}_a}{dt} = \mathbf{v}_a.$$

(9)

When dealing with liquids and elastic solids it is better to replace (9) by the XSPH variant (Monaghan[21]). In this case (9) is replaced by

$$\frac{d\mathbf{r}_a}{dt} = \mathbf{v}_a + \epsilon \sum_b m_b \frac{\mathbf{v}_{ba}}{\bar{\rho}_{ab}} W_{ab},$$

(10)

where $\bar{\rho}_{ab}$ denotes $(\rho_a + \rho_b)/2$ and a good value for ϵ is 0.5. If the position is changed according to this rule then linear and angular momentum are still conserved. It is interesting that the use of this average velocity is the first step on the road to implementing the alpha turbulence model (for references see Holm *et al.*[22]).

The above equations form a set of ordinary differential equations. To complete the system the pressure P must be specified in terms of the thermal energy and density. Other terms involving body forces, heat conduction and heat sources can be included as needed.

2 Applications to astrophysics

The first applications of SPH were to gas dynamics in astrophysics and because there is extensive use of SPH for these problems I will only briefly mention some key points. The primary advantages of SPH for astrophysical problems are that (a) it requires no special symmetry, (b) the resolution automatically changes with the density of the material, (c) the computing effort is concentrated where the material is, and (d) the method integrates perfectly with tree codes for the calculation of self gravity because tree codes are based on particles[23]. Finally, once you have written the SPH code, it is usually easy to include more complicated physics. For a review of astrophysical applications see Benz[3].

3 Applications to liquids

A very large number of environmental and industrial problems require the study of nearly incompressible fluids such as water. The standard finite difference technique for treating these problems is to use an implicit method where the pressure is determined from Poisson's equation. However, no entirely satisfactory implicit SPH methods are known for this problem, though Cummins and Rudman[24] have made significant improvements using a projection technique.

A simple explicit method with an artificial equation of state sufficiently stiff to guarantee that the density fluctuations are small can be used to treat a wide range of problems (Monaghan[9]) . Since $\delta\rho/\rho \sim M^2$ where M is the Mach number, we only need to ensure that $M \sim 0.1$. The following equation of state has been found satisfactory

$$P = \frac{100\rho_o V_o^2}{\gamma} \left[\left(\frac{\rho}{\rho_o} \right)^\gamma - 1 \right], \tag{11}$$

where V_o is the maximum velocity expected in the motion and $\gamma \sim 7$. The speed of sound is then $10V_o$. The price paid is that the time step is controlled by the Courant condition rather than a time step determined from the much smaller bulk speed of the material. In most cases V_o can be determined easily. In some cases it may be difficult. In the worst case it may be necessary to run the calculation again with the factor 100 replaced by 200 to ensure that the results are not artifacts.

4   Boundaries

Astrophysical problems do not have solid boundaries though they do have interfaces. Industrial and environmental problems have boundaries which may be rigid, moving, erodable, porous or some combination. In industrial problems the boundaries may have a simple geometry though the surfaces of car engines, for example, are highly complicated. In the environment the geometry of the boundaries is usually complicated.

Simple reflecting boundaries can be handled by reflecting a layer of SPH particles about the boundary and reversing the normal component of the velocity. The layer needs to be at least as wide as the range of the kernel. Periodic boundaries can be treated in a similar way. More complicated boundaries can be simulated by using Peskin's imbedded boundary method[25]. In this method the boundary is replaced by a set of particles which interact with the fluid through forces. Peskin used a finite difference method for the fluid and imbedded his particles in it. However, from the perspective of SPH, the boundary particles are just another set of particles and they fit naturally into the SPH method. The boundary particles can have a specified motion or react to forces when they form part of a moving rigid body.

The implementation of this boundary method is described by Monaghan[26] and by Monaghan and Kos[10].

An alternative approach to fixed boundaries which shows a great deal of promise is the application of projection methods by Cummins and Rudman[24].

5   Rigid bodies and fluids

The boundary particles can define a rigid body of arbitrary shape and the forces and torques on the rigid body can be calculated from the forces on these boundary particles. The equation of motion of the center of mass of a rigid body of mass M with center of mass at $\mathbf{R}$ and velocity $\mathbf{V}$ is (we consider just the two dimensional case in the following)

$$M\frac{d\mathbf{V}}{dt} = \mathbf{F},\tag{12}$$

where $\mathbf{F}$ is the total force on the body. The equation for the angular velocity Ω about the center of mass is

$$I\frac{d\Omega}{dt} = \tau,\tag{13}$$

where I is the moment of inertia (a scalar for the present case). The boundary particles defining the box can be distinguished from those defining the fixed boundary by giving them an integer tag. Denoting the force per unit mass on rigid body boundary particle k by $\mathbf{f}_k$ the above equations become

$$M\frac{d\mathbf{V}}{dt} = \sum_k m_k \mathbf{f}_k, \qquad (14)$$

and the equation for the angular velocity Ω about the center of mass is

$$I\frac{d\Omega}{dt} = \sum_k m_k (\mathbf{r}_k - \mathbf{R}) \times \mathbf{f}_k. \qquad (15)$$

The rigid body boundary particles move as part of the rigid body so that the position of boundary particle k is given by

$$\frac{d\mathbf{r}_k}{dt} = \mathbf{V} + \Omega \times (\mathbf{r}_k - \mathbf{R}). \qquad (16)$$

The force per unit mass $\mathbf{f}_k$ on the boundary particle k is due to the fluid particles unless the rigid body strikes the wall. Neglecting the latter case this force can be written as a sum over the fluid particles

$$\mathbf{f}_k = \sum_a \mathbf{f}_{ka}, \qquad (17)$$

where $\mathbf{f}_{ka}$ denotes the force per unit mass on k due to a. The precise form of this force depends on how the boundary force is implemented. However, to ensure that linear and angular momentum are conserved, the force on a due to k must be opposite in direction to that on k due to a. If the perpendicular distance from the boundary particle to the fluid particle is denoted by y and the tangential distance by x then a suitable form for the force per unit mass is

$$\mathbf{f}_{ka} = \frac{m_a}{m_a + m_k} B(x, y)\mathbf{n}_k, \qquad (18)$$

where $B(x, y)$ denotes the function associated with Peskin's boundary method. Similarly the force per unit mass on fluid particle a due to boundary particle k is

$$\mathbf{f}_{ak} = -\frac{m_k}{m_a + m_k} B(x, y)\mathbf{n}_k. \qquad (19)$$

The conservation of total linear momentum follows directly. The conservation of angular momentum is a little more subtle and depends on the interpolation.

These equations can be integrated along with the equations for the fluid particles. If a very heavy rigid body strikes a fixed boundary then the boundary force must be strong enough to repel the solid body.

6　Multiphase and multimaterial

Systems with solid, liquid and gas phases interacting are common. A simple case is the flow of more than one fluid as occurs, for example, when a river carrying nearly fresh water meets the ocean. In this case each fluid can be represented by its own set of SPH particles with appropriate masses and appropriate ρ_o in the equation of state (11). Interfaces, as mentioned earlier, do not need special treatment. For an application to gravity currents see Monaghan et al.[12].

More complicated examples involve solid and fluid phases. Consider for example a river carrying solid grains, or a pyroclastic outburst from a volcano where ash, rock and hot gas form the flow. This problem can be set up easily using the SPH implementation of the equations considered by Valentine and Wohletz[27]. A simpler problem is the flow of similar granular particles down a slope into a body of water. This is a very simplified model of a debris flow into a lake or into the sea. If the granules are very small compared to the characteristic dimensions of the flow they can be treated as a fluid (see for example Harlow and Amsden[28]). Unfortunately the bulk properties (viscosity, energy loss, etc.) of granular material are poorly understood but, given any formulation of the bulk properties of the fluid, it can be replaced by a set of SPH particles. One form of the SPH equations has been given by Monaghan and Kocharyan for dusty gas[29]. In our Epsilon laboratory we are studying small scale granular flows into water by a combination of experiment and simulation to understand the physics of wave production. Our ultimate aim is to predict the generation of tsunamis by the flank collapse of island volcanoes. The SPH formulation allows a wide range of interactions to be included very easily.

7　Elastic fracture

Fracture arises from the growth of flaws in the material. Benz and Asphaug[1,13] began studying this problem because of questions concerning the growth of planetesimals in the solar system. The problem is a natural one for SPH because the material is transformed from a coherent solid into a very large number of small fragments. It is difficult to treat this transformation accurately using a finite difference or finite element scheme. It is also difficult for finite difference codes to follow the damage in elements of the material. Benz and Asphaug found that an SPH code gave good predictions of the fracturing and the distribution of fragment sizes.

The SPH acceleration equation takes the form

$$\frac{dv_a^i}{dt} = \sum_b m_b \left(\frac{\sigma_a^{ij}}{\rho_a^2} + \frac{\sigma_b^{ij}}{\rho_b^2} + \Pi_{ab} \right) \nabla_a^i W_{ab},$$

(20)

where the stress is now defined by

$$\sigma^{ij} = -P\delta^{ij} + S^{ij},$$

(21)

and S^{ij} is the deviatoric stress. This deviatoric stress is determined by another differential equation based on Hooke's law (Benz and Asphaug[1]). Fracture and fragmentation can be determined by following the damage of the particles (Benz and Asphaug[1]). This is a further ordinary differential equation. In this problem the SPH particles, in addition to carrying the properties mentioned earlier, also carry the components of the stress tensor and the damage.

A problem with the early SPH codes for this problem is that, when applied to elastic solids, an instability occurred (Swegle et al.[30]). The instability arose when the material was in tension and the result was artificial clumping of the material. Unfortunately this clumping obscures the fragmentation in brittle impact although, in the calculation of Benz and Asphaug, the fragmentation was initiated well before the tensile instability began. In recent work it has been shown how the tensile instability can be prevented (Monaghan[15]) and applications to a wide variety of problems (Gray et al.[31]) show that SPH works exceptionally well for elastic problems. The instability is prevented by including a small repulsive term in the acceleration equation.

8 Summary

In a recent book Freeman Dyson[32] remarks how the advance of science depends as much on new techniques as it does on grand theory. SPH is one of those techniques which gives us the power to experiment with and explore the application of physical theory to the phenomena of nature. In this paper I have only had the space to mention some of the possible applications that can be forseen. There are many extensions of the multiphase and multimaterial applications possible especially in the earth sciences. One fascinating class of problems involves freezing of binary systems which are important in the mineral processing industry and in the discussion of sea ice. The salty ocean produces ice which is nearly pure water, and the increase in salinity of the sea water at the surface produces convection. Alloy freezing has other interesting properties. Another class of problems arise in the treatment of ultra relativistic heavy ion collisions using hydrodynamics. This is like the planetesimal collision problem in that many fragments are produced and one wants a natural transition from a fluid to particles. SPH handles that task effortlessly.

References

1. W. Benz and E. Asphaug, *Icarus* **107**, 98 (1994)
2. J. J. Monaghan, *Ann. Rev. Astron. Astro.* **30**, 543 (1992).
3. W. Benz in *The Numerical Modelling of Nonlinear Stellar Pulsation*, J. R. Buchler, ed., (Kluwer, Utrecht, 1990).
4. T. Belytschko, Y. Krongauz, D. Organ, M. Fleming and P. Krysl, *Comp. Meth. App. Eng.* **139**, 3 (1996).
5. F. H. Harlow, *J. Assoc. Comp. Mach.* **4**, 137 (1957)
6. J. Murray, J. J. Monaghan, S. Maddison, *Publ. Astron. Soc. Pacific* **55**, 13 (1995)
7. J. J. Monaghan, *J. Comp. Phys.* **136**, 298 (1997)
8. E. Chow and J. J. Monaghan, *J. Comp. Phys.* **62**, 183 (1997).
9. J. J. Monaghan, *J. Comp. Phys.* **110**, 399 (1994)
10. J. J. Monaghan and A. Kos, *J. Ports Waterways and Coastal Eng.* **125**, 145 (1999).
11. J. J. Monaghan and A. Kos, *Physics of Fluids* **12**, 622 (2000).
12. J. J. Monaghan, R. F. Cas, A. Kos and M. Hallworth, *J. Fluid Mech.* **379**, 39 (1999).
13. W. Benz and E. Asphaug, *Comp. Phys. Comm.* **87**, 253 (1995)
14. P. Randles and L. D. Libersky, *Comp. Method. App. Mech. Eng.* **139**, 375 (1996)
15. J. J. Monaghan, *J. Comp. Phys.* **159**, 290 (2000)
16. Y. Zhu, P. J. Fox and J. P. Morris, *J. Comp. Phys.* **116**, 123 (1995).
17. M. Steinmetz and E. Mueller, *Astron. and Astroph.* **268**, 391 (1992).
18. J. J. Monaghan and R. A. Gingold, *J. Comp. Phys.* **52**, 374 (1983).
19. P. W. Cleary, *New Implementations of Viscosity: Tests with Couette Flow* (Tech. Notes CSIRO, Tech. Rep. DMS-C96/32, 1996).
20. P. W. Cleary and J. J. Monaghan, *J. Comp. Phys.* **62**, 183 (1997).
21. J. J. Monaghan, *J. Comp. Phys.* **82**, 1 (1989)
22. D. D. Holm, *Physica D* **133**, 215 (1999).
23. J. Barnes and P. Hut, *Nature* **324**, 446 (1988).
24. S. Cummins and M. Rudman, *J. Comp. Phys.* **152**, 584 (1999).
25. C. S. Peskin, *J. Comp. Phys.* **25**, 220 (1977).
26. J. J. Monaghan, *Improved modelling of boundaries*, App. Maths. Rep. 95/30, Monash University, Unpublished.
27. G. A. Valentine and K. W. Wohletz, *J. Geophys. Res.* **94**, 1867 (1989).
28. F. H. Harlow and A. A. Amsden *J. Comp. Phys.* **17**, 19 (1975).
29. J. J. Monaghan and A. Kocharyan, *Comp. Phys. Comm.* **87**, 225 (1995)
30. J. Swegle, D. L. Hicks and S. W. Attaway, *J. Comp. Phys.* **116**, 123 (1995).

31. J. P. Gray, J. J. Monaghan and R. Swift, *SPH Elastic Dynamics*, submitted, *Comp. Method. App. Mech. Eng.* (2000).
32. F. J. Dyson, *The Sun, The Genome and the Internet*, Oxford University Press, Oxford.

A PARALLEL LATTICE-BOLTZMANN METHOD FOR LARGE SCALE SIMULATIONS OF COMPLEX FLUIDS

MAZIAR NEKOVEE, JONATHAN CHIN, NÉLIDO GONZÁLEZ-SEGREDO
AND PETER V. COVENEY

Centre for Computational Science,
Department of Chemistry, Queen Mary, University of London,
Mile End Road, London E1 4NS, U.K.
E-mail: m.nekovee@qmw.ac.uk, p.v.coveney@qmw.ac.uk

The lattice-Boltzmann method is a relatively new approach for the simulation of complex fluids which has a mesoscopic character, intermediate between continuum fluid dynamics approach and the atomistic approach based on molecular dynamics. Major advantages of the method include its ability to deal with the complex dynamics of interfaces and complicated boundaries, and the inherently parallel nature of the underlying algorithm. We review a novel lattice-Boltzmann scheme for simulation of binary and ternary fluids involving surfactant[1,2], discuss a parallel implementation of the algorithm and present preliminary results of large-scale simulations of the complex dynamics of ternary amphiphilic fluids performed on massively parallel platforms.

1 Introduction

The lattice-Boltzmann method (LB) is a relatively new approach in computational fluid dynamics. It originated from the lattice-gas automata (LGA) model and has a mesoscopic character, as opposed to the conventional continuum approach based on numerical solution of the Navier–Stokes equations and the microscopic approach based on molecular dynamics. The key idea behind the method is to model fluid flows by simplified kinetic equations that describe the time evolution of distribution function of particles having a discrete set of velocities and moving on a regular lattice[3]. The computationally demanding tracking of individual molecules is thus avoided and at the same time macroscopic or hydrodynamic effects naturally emerge from mesoscale lattice-Boltzmann dynamics, provided that the LB equations possesses the correct and necessary conservation laws and symmetries[3]. The LB method has found many applications in fluid dynamical problems in which more conventional continuum approaches face difficulties, such as solid–fluid suspensions, multiphase and multicomponent flow, flow in porous media and reaction diffusion systems[3]. A promising area of application of lattice-Boltzmann is for modeling amphiphilic fluids which consist of two immiscible phases (such as oil and water), together with an amphiphile (surfactant) species, such as detergent. The study of these fluids is of relevance to a wide variety of industrial, chemical and biological applications and is of great fundamental interest but their nonequilibrium dynamics and hydrodynamics are difficult to simulate using continuum approaches based on Navier–Stokes equations. Very recently we have developed a lattice-Boltzmann model for simulation of amphiphilic fluids[1,2]. The key features of our model are that the orientational degrees of freedom of surfactant molecules are included in the dynamics of LB equations and the coupling between fluid components is realized from bottom to top by introducing self-consistent mean-field forces between fluid particles. These features distinguish our approach from previous LB models of amphiphilic fluids[4] and are critical in describing most of the complex phenomenology of amphiphilic systems. The simplified treatment of surfactant molecules in our model means that chemical specificity cannot be fully taken into account. Atomistic approaches based on molecular dynamics which deal with "real" surfactant molecules are chemically specific but are still computationally too demanding to access large time dynamics involved in many problems of interest, such as self-assembly, which are easily accessible to our LB simulations. Very recently, we have pioneered a more fully bottom-up approach that applies systematic coarse-graining to an underlying many-body, molecular dynamics description, and results in a multiscale dissipative dynamics method[5]. Its numerical implementation, and application to complex fluids, are currently both in their infancy, however.

Both CPU time and memory requirements of our algorithm scale as $M \times N^D$ where M is the number of discrete velocity vectors, N is the linear size of the system and D is the spatial dimension. In two dimensions it is possible to study adequate system sizes and time scales using typical workstations. In three dimensions, however, the serial algorithm quickly becomes prohibitive in terms of computer memory and CPU time for moderately sized systems. Fortunately, an important feature of the LB method is its intrinsically parallel structure and we have implemented a parallel version of our algorithm which allows us to perform very large-scale $3D$ simulations on massively parallel platforms. In this paper we give a brief overview of our LB scheme for amphiphilic fluids, describe a parallel implementation of the algorithm and examine its performance on massively parallel platforms. We present preliminary results of parallel LB simulations of phase separation and self-assembly in binary and ternary systems and conclude with an outlook on future applications of our method to complex fluids.

2 Lattice-Boltzmann algorithm for amphiphilic fluids

In our scheme nonamphiphilic molecules, such as oil and water are modeled as point particles while amphiphilic molecules are assumed to carry a dipole vector whose orientation can vary continuously in time. The lattice-Boltzmann equations describing the dynamics of a fluid comprising of two immiscible nonamphiphilic components (e.g., oil and water) and an amphiphile are[2]

$$f_i^\sigma(\mathbf{x}+\mathbf{c}_i\Delta t, t+\Delta t) - f_i^\sigma(\mathbf{x},t) = -\Delta t \frac{f_i^\sigma - f_i^{\sigma(eq)}}{\lambda_\sigma} + \sum_{\bar\sigma}\sum_j \Lambda_{ij}^{\sigma\bar\sigma} f_j^{\bar\sigma} + \sum_j \Lambda_{ij}^{\sigma s} f_j^s \tag{1}$$

$$f_i^s(\mathbf{x}+\mathbf{c}_i\Delta t, t+\Delta t) - f_i^s(\mathbf{x},t) = -\Delta t \frac{f_i^s - f_i^{s(eq)}}{\lambda_s} + \sum_{\sigma}\sum_j \Lambda_{ij}^{s\sigma} f_j^{\sigma} + \sum_j \Lambda_{ij}^{ss} f_j^s \tag{2}$$

$$\mathbf{d}(\mathbf{x},t+\Delta t) - \bar{\mathbf{d}}(\mathbf{x},t) = -\Delta t \frac{\mathbf{d}(\mathbf{x},t) - \mathbf{d}^{(eq)}(\mathbf{x},t)}{\lambda_d}. \tag{3}$$

The first two equations describe the time evolution of discrete velocity distribution functions f_i^σ and f_i^s belonging to component σ (σ = oil, water) and surfactant (s), respectively and the third equation describes the time evolution of surfactant dipoles $\mathbf{d}(x,t)$. In the above equations $\mathbf{c}_i$ ($i = 0, 1 \ldots M$) are a set of discrete velocity vectors, $\mathbf{x}$ is a point on the underlying spatial grid and Δt is the timestep. Furthermore, $f_i^{\sigma(eq)}$, $f_i^{s(eq)}$ and $\mathbf{d}^{(eq)}$ are local equilibrium distribution functions, λ_σ, λ_s and λ_d are relaxation times and $\bar{d}$ is the average dipole at site $\mathbf{x}$ prior to collisions. The first term in the right-hand sides of Eqs. (1) and (2) is the standard BGK collision operator[3]. The terms $\Lambda_{ij}^{\sigma\bar\sigma}$, $\Lambda_{ij}^{s\sigma}$, $\Lambda_{ij}^{s\sigma}$ and Λ_{ij}^{ss} are matrix elements of collision operators which result from mean-field interactions among different fluid components[2]. These cross-collision terms give rise to phase separation in binary oil–water systems and microemulsion and self-assembly in ternary amphiphilic mixtures. Hydrodynamic quantities such as number densities $n^{\sigma,s}$ and velocities $\mathbf{u}^{\sigma,s}$ of each fluid component are obtained from velocity moments of the corresponding distribution functions. Kinematic viscosities of each fluid component is controlled by the corresponding relaxation time λ_σ and λ_s. The mean-field force between nonamphiphilic particles is obtained from[1]

$$\mathbf{F}^{\sigma\bar\sigma}(\mathbf{x},t) = -n^\sigma(\mathbf{x},t) \sum_{\bar\sigma}\sum_{\mathbf{x}'} g_{\sigma\bar\sigma} n^{\bar\sigma}(\mathbf{x}+\mathbf{c}_i\Delta t, t)\mathbf{c}_i \tag{4}$$

where $g_{\sigma\bar\sigma}$ (> 0 for immiscible fluids) is a force coupling constant, whose magnitude controls interfacial tension.

The forces acting among amphiphilic molecules themselves and between amphiphilic molecules and oil/water molecules depend not only on the distance between particles but also on the orientation of amphiphilic dipoles. For example the force acting on oil or water particles by amphiphiles is[1]

$$\mathbf{F}^{\sigma,s}(\mathbf{x},t) = -2n^{\sigma}(\mathbf{x},t)g_{\sigma s}\sum_{i\neq 0}\mathbf{d}(\mathbf{x}+\mathbf{c}_i\Delta t,t)\cdot(\mathbf{I}-\frac{\mathbf{c}_i\mathbf{c}_i}{c_i^2}D)n^s(\mathbf{x}+\mathbf{c}_i\Delta t,t),\quad (5)$$

where $\mathbf{I}$ is the second-rank unit tensor.

3 Parallelization

Each timestep of our LB algorithm consists of the following substeps:
Advection: The particles are propagated along their velocity vectors to adjacent sites.
Force calculation: At each site, the total force on each particle species is calculated using Eqs. (4), (5), etc.
Collisions and update: The equilibrium densities for each species for each velocity vector are found and the particle distribution and surfactant dipoles are updated according to Eqs. (1)–(3).

Parallelization was performed utilizing the single program multiple data (SPMD) model, wherein the same program is executed on all processors and by means of a domain decomposition strategy. This strategy is very suitable for grid-based and semilocal algorithms like lattice-Boltzmann, finite-difference and finite-element methods. The underlying 3D lattice is partitioned into subdomains and each subdomain is assigned to one processor. Such a decomposition may be done in one dimension (slices), two dimensions (rods) or three dimensions (boxes). We have used box decomposition, which offers more flexibility and also results in a minimum surface/volume ratio of subdomains, hence minimizing communication overheads. Each processor is responsible for the particles within its subdomain and performs exactly the same calculations on these particles. Two rounds of communication between neighboring subdomains are required: at the propagation step, where particles on a border node can move to a lattice point in the subdomain of a neighboring processor, and in evaluating the forces from Eqs. (4), (5), etc. By using a ghost layer of lattice points around each subdomain, the propagation and collision steps can be isolated from the communication step. Before the propagation step is carried out the values at the border grid points are sent to the ghost layers of the neighboring processor and after the propagation step an additional round of communication is performed to update the ghost layers. Communication between processors is performed using the message passing standard MPI.

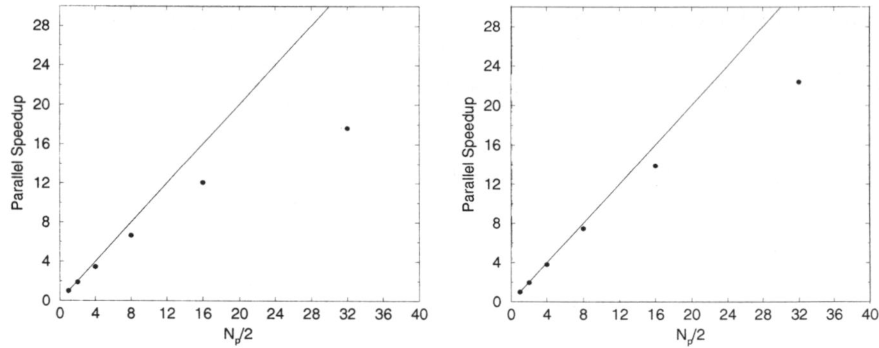

Figure 1. Parallel speedup measured on a CRAY T3E 1200E. Left panel shows speedup of the binary code. Right panel shows speedup of the ternary amphiphilic code. Filled circles are the computed results; the solid line shows the linear speedup.

The introduction of surfactant interactions and dynamics significantly increases both memory requirements and execution time of the algorithm and for this reason we implemented two parallel versions of the algorithm, one for binary immiscible systems and one for ternary amphiphilic systems. Parallel performance of both codes was benchmarked on the distributed memory CRAY T3E 1200E massively parallel platform. Preliminary benchmarks on SGI Origin 2000 were also performed which showed that the code runs with high parallel efficiency on this platform, whose shared memory architecture is completely different from the T3E. All simulations were performed on 64^3 systems using 19 velocity vectors for each component and 500 timesteps. Fluid densities at each lattice site, plus surfactant dipole moments in case of ternary calculations, were written out periodically at every 50 timesteps of the simulations. We use parallel speedup—the ratio of the execution time of the parallel code on the minimum possible number of processors N_p^0 (2 in our case since the code does not fit in the memory of a single processor) to the execution time of the same code on N_p processors—to measure the efficiency of our parallel algorithm. LB simulations using the binary code took 4500 CPU seconds on 2 processors of a CRAY T3E 1200E (Dec Alpha EV6 600 MHZ processors with 256 Mbytes of memory each). Figure 1 (left panel) shows the plot of parallel speedup versus number of processors of the binary code. Also shown is the linear speedup curve, which would be obtained if there were no communications between processors and the computation time per one lattice-Boltzmann grid update was independent of the number of processors. As can be seen the speedup is close to linear up to 16 processors after which it starts to tail off due to increase in communication overhead. Simulations using the ternary code took 11560 CPU seconds on 2 processors of a CRAY T3E 1200E. The parallel speedup of the ternary code is shown in Fig. 1 (right panel) together with linear speedup. As can be seen from this figure, parallel performance of the ternary code is better than that of the binary code and stays close to linear up to 32 processors. We believe this behavior to be a memory effect, resulting from the division of large arrays in the ternary code into smaller subarrays on each processor which then better fit into the cache of each CPU when $N_p > 2$. This reduces the time that each CPU spends in fetching data from the main memory, hence reducing the computational time per grid point for $N_p > 2$.

4 Phase separation and self-assembly in binary and ternary fluids

Below a critical temperature (spinodal point) oil and water are immiscible and do not mix. The dynamics of phase separation in an initially homogeneous binary mixture of oil and water has been extensively studied in two dimensions and less in three dimensions[6,7]. Figure 2 shows snapshots of our parallel LB simulations of phase separation of an initially homogenized 1 : 1 oil water mixture. The system size is 64^3 and the simulations were performed on 32 nodes of the T3E 1200E. Snapshots of this simulations are displayed in Fig. 2 and show spontaneous formation of small domains and their growth until the system reaches a completely separated state of, essentially, two distinct layers of fluid. A quantitative analysis of the dynamics of phase separation can be obtained by calculating the average domain size and analyzing its scaling behavior in certain asymptotic regimes. Our recent large-scale simulations using a serial version of our LB code demonstrated the ability of our model to give a correct quantitative description of phase separation in two dimensions[2] and we plan to use our parallel code to perform similar quantitative studies in 3D, using adequate system sizes.

The addition of surfactant to a binary oil–water mixture can arrest the growth of domains and result in self-assembly of complex structures on a mesoscopic length scale. These structures include micelles, emulsion droplets, sponge phases as well as structures with long-range order such as lamellar, cubic and hexagonal phases[8]. In Fig. 3 we show an example of our parallel lattice-Boltzmann simulations of self-assembly in a ternary mixture of oil, wa-

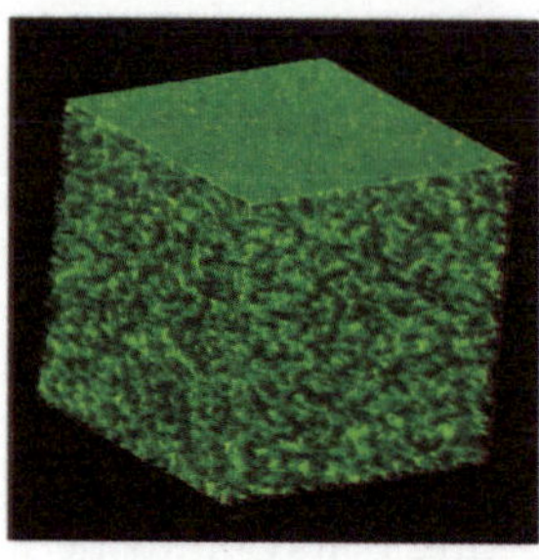 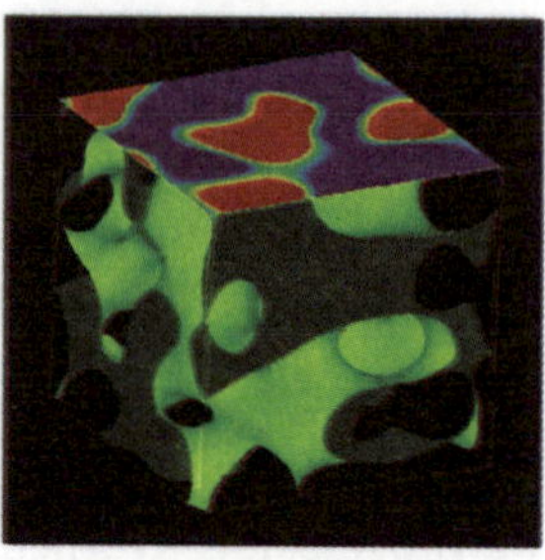 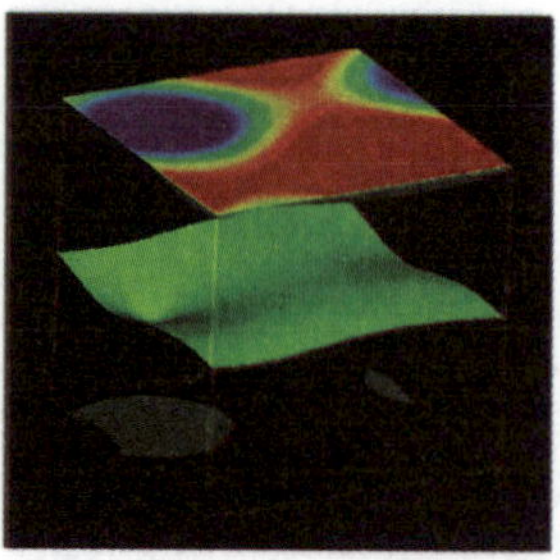

Figure 2. Snapshots from a parallel LB simulation of phase separation in a binary oil–water mixture . The average density of both components is 0.5. From left to right timesteps 0, 1000 and 5000 of simulations are shown. In all snapshots the isosurface shows the interface between oil and water. The red (blue) coloring on the slice planes indicates high oil (water) concentration and the green coloring (roughly) indicates interfaces. System size is 64^3 and the simulations were performed on 32 nodes of a CRAY T3E 1200E massively parallel supercomputer.

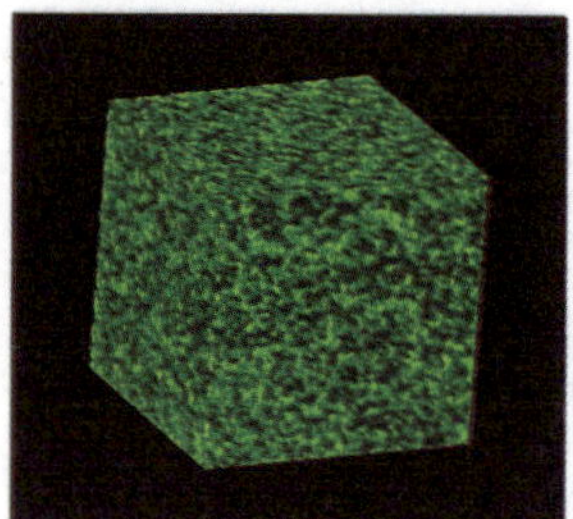 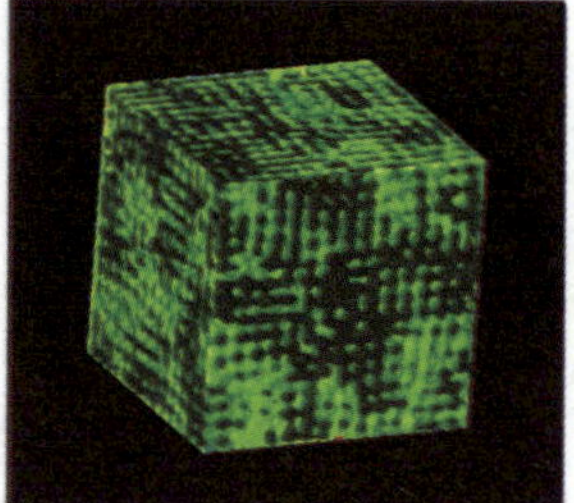 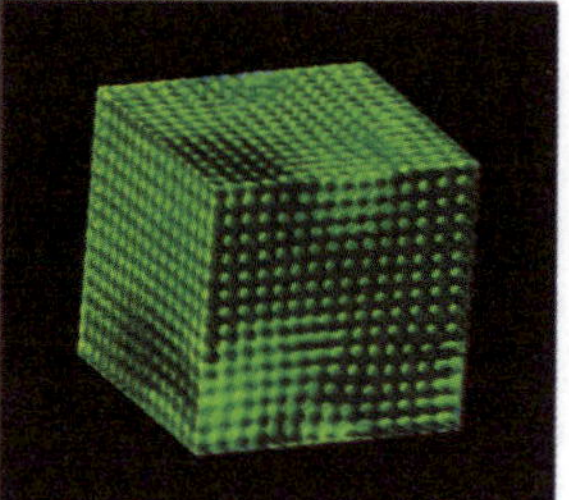

Figure 3. Snapshot of parallel LB simulations of self-assembly in a ternary amphiphilic fluid consisting of oil, water and surfactant. The average densities of oil, water and surfactant are 0.2, 0.2 and 0.05, respectively. From left to right timesteps 0, 100 and 2000 of the simulations are shown. In all snapshots the isosurface shows the interface between oil and water. System size is 64^3 and the simulations were performed on 64 nodes of a CRAY T3E 1200E.

ter and surfactant. The simulations were performed for a 64^3 system on 64 nodes of a CRAY T3E 1200E and took 35 minutes to reach 2000 timesteps. As can be seen from this figure, starting from an initial homogeneous mixture of oil, water and surfactant, the ternary amphiphilic fluid organizes itself in a perfectly ordered cubic structure. The existence of such ordered equilibrium phases in amphiphilic fluids is well-known[8]. However, the dynamics of self-assembly of such structures in three dimensions has been beyond the reach of mesoscale simulations until now (see also Boghosian *et al.*[7]). We stress that the phenomenologies calculated here are extremely difficult to simulate using conventional computational fluid dynamics approaches (based on numerical solution of Navier–Stokes equations). One major numerical difficulty these methods face is tracking the complicated fluid interfaces which undergo topological changes[9]; another more fundamental problem is that it is far from clear how to formulate a hydrodynamic description of such systems.

5 Outlook

Parallelization combined with the use of massively parallel computers has brought large-scale LB studies of amphiphilic fluids within our reach and we are currently in the process of performing extensive studies of the nonequilibrium dynamics of these systems. Other important applications of our parallel algorithm, currently underway, include the study of multiphase and multicomponent fluid flow through porous media. The lattice-Boltzmann method is particularly suited to such studies because of the ease with which complex boundaries can be implemented in this scheme. The high parallel efficiency of our code should allow us to perform these studies using realistic and accurate descriptions of porous media.

Acknowledgments

This work was partially supported by EPSRC under Grant No GR/M56234. Our calculations were performed on the CRAY T3E 1200E at Computing Service for Academic Research (CSAR) in Manchester.

References

1. H. Chen, B. M. Boghosian, P. V. Coveney and M. Nekovee, *Proc. Roy. Soc. London* A **456**, 2043 (2000).
2. M. Nekovee, P. V. Coveney, H. Chen and B. M. Boghosian, *Phys. Rev. E* **62** (2000).
3. For a recent review see S. Chen and G. D. Doolen, *Ann. Rev. Fluid. Mech.* **30**, 329 (1998).
4. A. Lamura, G. Gonnella and J. M. Yeomans, *Europhys. Lett.* **45**, 314 (1999); O. Theissen, G. Gompper and D. M. Kroll, *Europhys. Lett.* **42**, 419 (1998).
5. E. G. Flekkøy and P. V. Coveney, *Phys. Rev. Lett.* **83**, 1775 (1999); E. G. Flekkøy, P. V. Coveney and G. De Fabritiis, *Phys. Rev. E* **62**, 2140, (2000).
6. S. I. Jury, P. Bladon, S. Krishna and M. E. Cates, *Phys. Rev. E* **59**, R2535 (1999).

7. B. M. Boghosian, P. V. Coveney and P. Love, *Proc. Roy. Soc. London* A **456**, 1431 (2000).
8. G. Gompper and M. Shick, *Phase Transitions Crit. Phenomena* **16**, 1 (1994).
9. K. A. Smith, F. J. Solis, L. Tao, K. Thornton and M. Olvera de la Cruz, *Phys. Rev. Lett.* **91**, 84 (2000).

NATURAL CONVECTION IN A SLENDER CONTAINER

LUIS M. DE LA CRUZ

*Dirección General de Servicios de Cómputo Académico, Universidad Nacional
Autónoma de México, Circuito Exterior, Cd. Universitaria, 04510 México, D.F.,
México*
E-mail: lmd@labvis.unam.mx

EDUARDO RAMOS

*Centro de Investigación en Energía, Universidad Nacional Autónoma de México,
Apdo. Postal 34, 62540 Temixco, Morelos, México*
E-mail: eramos@servidor.unam.mx

Natural convection in a slender cylindrical cavity with square cross section and
aspect ratio (side to height) of 0.5 has been studied with the aim of understanding
the structure of the multiplicity of steady-state solutions and its transit to unsteady
flow. Prandtl number is considered equal to 5. The numerical code used for the
analysis is based on the control volume method and the integration strategy is
similar to SIMPLE. A continuation method is used to determine that a single
branch of solutions (trivially symmetric solutions are considered the same solution)
is found for $Ra \leq 6 \times 10^5$; then, two branches are found for $6 \times 10^5 \leq Ra \leq 7.5 \times 10^5$.
Calculations with $Ra > 8 \times 10^5$ indicate that the system presents a bifurcation
leading to irregular oscillations.

1   Introduction

The main motivation of the present study is the understanding of transition natural convection flows in slender containers with application to crystal
growth with the Czochralski method. It is probable that turbulence in small
aspect ratio containers accepts a relative simple description due to the restricted range of possible wavelengths and the large influence of the walls. An
indication of the probable simplification is supplied by the studies of transition natural convective flows in constrained Rayleigh–Benard flows. Gollub
et al.[1] made experimental observations in a rectangular box of aspect ratios
of 3.5:2.1:1.0 and concluded that the transit from steady state to turbulent
flow as the Rayleigh number is increased is made via two subharmonic bifurcations, then an interval of Rayleigh numbers where a quasiperiodic flow
is observed and finally at even higher Rayleigh numbers, chaotic flow is encountered. Mukutmoni and Yang[2] obtained a numerical solution to the conservation equations for the natural convection in a rectangular box with the
same aspect ratio as Gollub et al.[1] They were able to reproduce the first
bifurcation and a similar sequence of bifurcations from those observed experi-

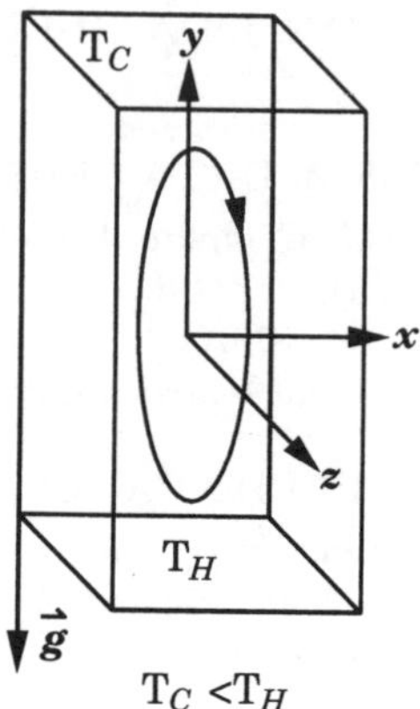

Figure 1. Geometry of the cavity and coordinate system.

mentally. Neumann[3] presented a numerical solution for natural convection in a cylindrical container and reported that for a cavity with height equal to its diameter $(h = d)$ the time dependent natural convection sets in with strongly nonharmonic fluctuations. Müller *et al.*[4] reported experimental observations of the Rayleigh number required for time-dependent oscillatory convection in cylindrical containers. The critical Rayleigh number for aspect ratio (diameter to height) 0.5 is approximately 8×10^6.

In the present study, we explore the bifurcation map for steady flows, i.e., flows obtained with Rayleigh numbers smaller than the critical; also, we describe some properties of unsteady flows obtained with Rayleigh numbers larger but close to the critical.

2 Model

The geometry of the cavity and the coordinate system used in the present study are shown in Fig. 1. The enclosure is assumed to be a rectangular parallelepiped of aspect ratios of 0.5 and 0.5 with top and bottom walls kept at constant cold (T_C^*) and hot (T_H^*) temperatures respectively; all other walls are adiabatic. The Boussinesq approximation is assumed to be valid.

The governing equations in terms of nondimensional variables are:

$$\nabla \cdot U = 0 \tag{1}$$

$$\frac{\partial u}{\partial t} + \nabla \cdot (uU) = -\frac{\partial p}{\partial x} + \Pr \nabla^2 u \tag{2}$$

$$\frac{\partial v}{\partial t} + \nabla \cdot (vU) = -\frac{\partial p}{\partial y} + \Pr \nabla^2 v + Ra \Pr T \tag{3}$$

$$\frac{\partial w}{\partial t} + \nabla \cdot (wU) = -\frac{\partial p}{\partial z} + \Pr \nabla^2 w \tag{4}$$

$$\frac{\partial T}{\partial t} + \nabla \cdot (TU) = \nabla^2 T \tag{5}$$

The spatial coordinates x, y, z are scaled with L, the cavity height and the time is scaled by L^2/α, where α is the thermal diffusion coefficient, $U = (u, v, w)$ is the velocity vector scaled with α/L, p is the pressure difference between the total pressure and the hydrostatic pressure due to gravity and is scaled by $\rho\alpha^2/L^2$, where ρ is the density. The nondimensional temperature (T) is defined by $(T^* - T_m)/\Delta T$, where T^* is the dimensional temperature of the fluid, $\Delta T = T_H^* - T_C^*$ and $T_m = (T_H^* + T_C^*)/2$.

Rayleigh (Ra) and Prandtl numbers (Pr) are defined by:

$$Ra = \frac{g\beta\Delta T L^3}{\nu\alpha},$$

and

$$Pr = \frac{\nu}{\alpha}$$

Here, g, β and ν are respectively the gravity acceleration, the thermal volumetric expansion coefficient and the kinematic viscosity. In all cases studied, the Prandtl number is 5. The boundary conditions required for the solution are:

$$x = -0.25, 0.25; \; -0.25 \le z \le 0.25, -0.5 \le y \le 0.5 \qquad u = v = w = 0 \qquad \frac{\partial T}{\partial x} = 0 \tag{6}$$

$$z = -0.25, 0.25; \; -0.25 \le x \le 0.25, -0.5 \le y \le 0.5 \qquad u = v = w = 0 \qquad \frac{\partial T}{\partial z} = 0 \tag{7}$$

$$y = -0.5, 0.5; \; -0.25 \le x \le 0.25, -0.25 \le z \le 0.25 \qquad u = v = w = 0 \qquad T = -y \tag{8}$$

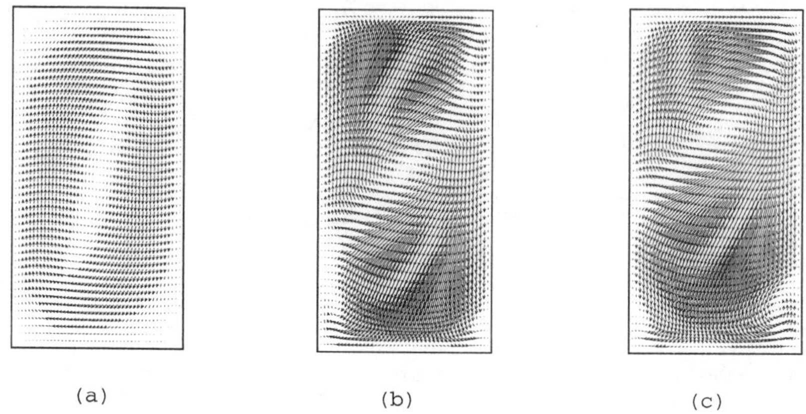

Figure 2. Velocity field for (a) $z = 0.0$ and $Ra = 10^5$, (b) $z = 0.0$ and $Ra = 7.5 \times 10^5$, branch 1, and (c) $z = 0.0$ and $Ra = 7.5 \times 10^5$, branch 2.

3 Numerical model

The numerical model used to integrate the conservation equations is essentially the same as that described in Mukutmoni and Yang [5]. The finite volume discretization scheme is used together with a version of the SIMPLE method for integrating balance equations. A quadratic approximation is used to discretize the convective terms as recommended by Leonard[6]; the backward explicit Euler method is used for the time integration. All calculations were made with regular mesh of 48^3 nodes. The step for both the pseudotime in the continuation procedure (see below) and the time in the unsteady calculations is 10^{-5} nondimensional time units.

4 Results

It is known that the possibility of multiple solutions of this kind of flows is generic and the determination of all possible branches and their stability in a properly defined space is the ultimate goal of a comprehensive description of these flows. The determination of a branch of solutions can be done using tools of parametric analysis like continuation[7]. In the present investigation, we use a predictor–corrector continuation strategy based on a pseudotime integration as follows: assume that the solution for Ra_1 is known. Then, modify the Rayleigh number (Ra_2 say) and perform a time-dependent integration until the steady state corresponding to Ra_2 is reached.

Before presenting the numerical results, it is convenient to point out some symmetries inherent to the physical conditions that must be considered when analyzing the multiplicity of solutions. Equations (1) to (3) and the boundary conditions are symmetric with respect to the following transformations:

$$
\begin{array}{lll}
x \rightarrow z & u \rightarrow w & p \rightarrow p \\
y \rightarrow y & v \rightarrow v & T \rightarrow T \\
z \rightarrow -x & w \rightarrow -u &
\end{array}
$$

and

$$
\begin{array}{lll}
x \rightarrow -x & u \rightarrow -u & p \rightarrow p \\
y \rightarrow y & v \rightarrow v & T \rightarrow T \\
z \rightarrow -z & w \rightarrow -w &
\end{array}
$$

These transformations represent a $\pi/2$ and a π rotation of the axis of coordinates around the vertical axis respectively. The multiplicity of solutions that arise from these symmetries are considered trivial in the sense that they do not modify the total heat transferred by the convective cell.

4.1 Steady flow

The onset of convection motion for this geometrical configuration takes place at $Ra = 1.6 \times 10^4$ (de la Cruz[8]). This is the first bifurcation point. For larger Rayleigh numbers stable or unstable convective solutions coexist with the unstable conductive solution.

At Rayleigh numbers smaller than approximately 4×10^5 the convective flow pattern is dominated by a practically bidimensional single cell with axis of rotation parallel to z axis. This structure fills most of the cavity. Figure 2(a) shows this flow for $Ra = 10^5$. The flow obtained at $Ra = 4 \times 10^5$ is also dominated by a single cell, but the convective cell develops a three dimensional structure. Using the continuation strategy it is possible to obtain solutions along this branch. It is observed that the basic flow undergoes quantitative but no qualitative changes, up to $Ra = 7.5 \times 10^5$.

At $Ra = 7.5 \times 10^5$, small perturbations to the steady state flow on branch 1, take the solution to a qualitatively different steady state flow. This convective pattern is a solution in branch 2. Figures 2(b) and (c) show the velocity field at $z = 0.0$ plane with $Ra = 7.5 \times 10^5$, for the two branches. The flows displayed are clearly different; notice that the central stagnation point is at a higher position in branch 2 and also a small vortex in the lower right-hand

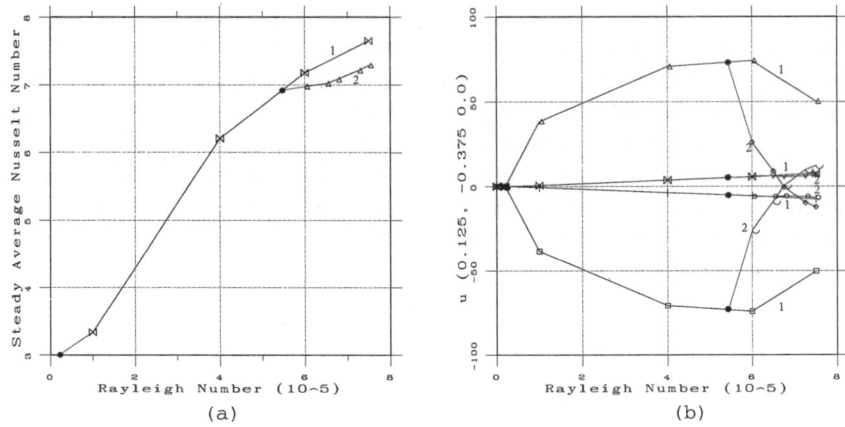

Figure 3. (a) Branching map I: Average Nusselt number as a function of Ra. (b) Branching map II: Horizontal velocity at point $(0.125, -0.375, 0.0)$ as a function of Ra. Branching points are denoted by •. The lines between the branching point and $Ra = 6 \times 10^5$ in branch 2 is extrapolated in both maps.

corner is present in the solution of branch 2. Applying the continuation strategy starting with $Ra = 7.5 \times 10^5$ and reducing the Rayleigh number, it is possible to construct the second branch with flows displaying small vortices.

Figure 3(a) shows the average Nusselt number for the various cases explored. Extrapolating, it can be asserted that the branching point is approximately located at $Ra = 5.5 \times 10^5$.

The choice of average Nusselt number as the measure in the branching map, hides trivial symmetries. In order to display all solutions of the system, Fig. 3(b) shows the branching map using the horizontal velocity component at the point $(0.125, -0.375, 0.0)$, including trivial symmetries as well.

Applying the continuation strategy to branch 2 it is found that flow at $Ra = 8.0 \times 10^5$ displays a critically damped oscillation with a frequency of approximately 20 oscillations per nondimensional time units. See Fig. 4(a). This indicates that the system undergoes a bifurcation at approximately this Rayleigh number. An analysis of the flow patterns obtained for the peaks and troughs of the oscillatory Nusselt number, makes it evident that the flow undergoes small, quantitative changes, but preserving the same qualitative pattern of one main convective cell shown in Fig. 2(c).

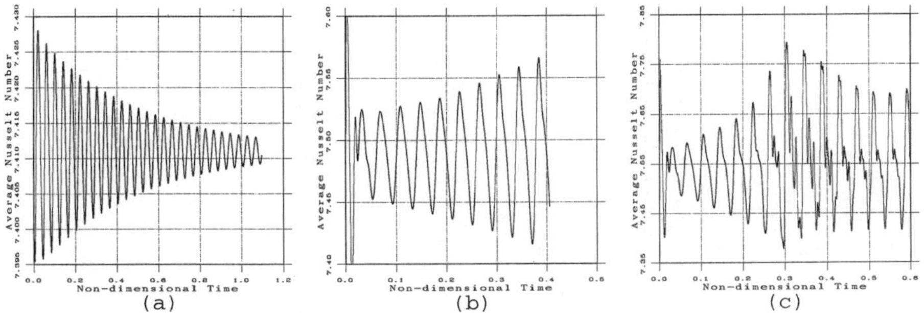

Figure 4. Nusselt number as a function of time for (a) $Ra = 8 \times 10^5$, (b) $Ra = 8.3 \times 10^5$ and (c) $Ra = 8.5 \times 10^5$.

4.2 Transient flow

An oscillatory flow with self-sustained oscillations with increasing amplitude and few Fourier components is found at $Ra > 8.3 \times 10^5$; see Fig. 4(b). At $Ra = 8.5 \times 10^5$ (Fig. 4(c)) flow oscillations become more complicated incorporating high frequency components. Also, low frequency modes make the flow quasiperiodic.

At even higher Rayleigh numbers, the flow becomes very irregular as can be appreciated from Fig. 5 where the Nusselt number is plotted as a function of time for $Ra = 10^6$. In the time interval explored, the average Nusselt number displays an irregular oscillation with intervals of 0.03 nondimensional time units of approximately constant average Nusselt number. The convective pattern in these intervals coincides qualitatively with that obtained in branch 2 for steady flows.

5 Discussion and conclusions

Transient natural convection in a slender cylinder has been analyzed with a numerical method. A continuation strategy based on a pseudotime-dependent integration has been used to explore two branches that appear near the critical Rayleigh number at which steady flow becomes time-dependent. Marginally unsteady flow displays a constant frequency oscillation, type of branching. Oscillatory flow at higher Ra is nonperiodical. Future work will include the use of test functions (Seydel[7]) for determining the position of the branching

220

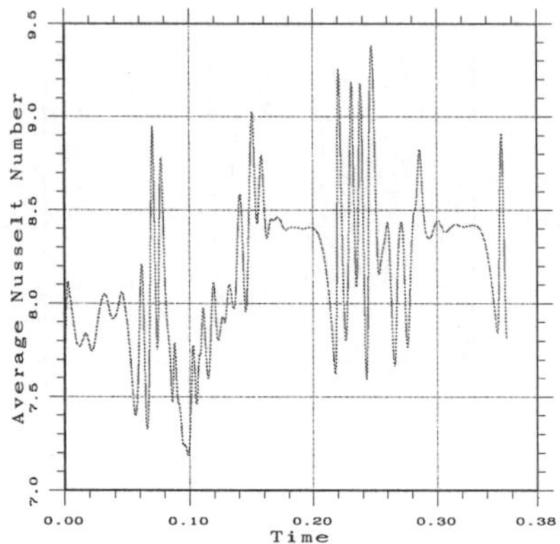

Figure 5. Nusselt number as a function of time for $Ra = 10^6$.

point with better accuracy.

Acknowledgments

Financial support from CONACyT through project number G0044E is gratefully acknowledged. Numerical calculations were performed on UNAM's SGI Origin 2000 at DGSCA.

References

1. J. P. Gollub, S. V. Benson and J. F. Steinman, in *Non linear dynamics*, H. G. Helleman, ed., 22–27 (New York Academy of Sciences, 1980).
2. D. Mukutmoni and K. T. Yang, *J. Heat Transfer* **115**, 360–366 (1993)
3. G. Neumann, *J. Fluid Mech.* **214**, 559–578 (1990)
4. G. Müller, G Neumann and W. Weber, *J. Crystal Growth* **70**, 78–93 (1984)
5. D. Mukutmoni and K. T. Yang, *J. Heat Transfer* **115**, 367–376 (1993)
6. B. P. Leonard, *Comp. Meth. Appl. Mech. Eng.* **19**, 59 (1979)
7. R. Seydel, *Practical bifurcation and stability analysis* (Springer, New York, 1994).
8. L. M. de la Cruz, *Natural convection in a rotating cavity* (in Spanish), B.Sc. Thesis, School of Sciences, UNAM, Mexico (1996).

EFFECT ON FLUID FLOW AND HEAT TRANSFER OF DEPHASING BETWEEN PLATES IN WAVY PLATE PASSAGES

R. ROMERO-MÉNDEZ, M. SOTO-CABRERO

Facultad de Ingeniería, Universidad Autónoma de San Luis Potosí
Edificio P, Zona Universitaria, 78290 San Luis Potosí, S.L.P., México

M. SEN, W. FRANCO-CONSUEGRA

Department of Aerospace and Mechanical Engineering, University of Notre Dame,
Notre Dame, IN, U.S.A.

A. HERNÁNDEZ-GUERRERO

FIMEE, Universidad de Guanajuato, Apdo. Postal 215-A, Salamanca, Gto.,
México

A finite element code is used to obtain information regarding the two-dimensional, steady-state flow field and heat transfer in a channel formed by a pair of equal amplitude, equal wave length, sinusoidal plates at different isothermal temperatures. A sufficiently long channel with several corrugations is considered so that periodic boundary conditions over one wave length are assumed in the flow inside the inner corrugations and the flow is assumed to be laminar. The nondimensional parameters governing the problem are similar to those encountered in compact heat exchangers. The primary objective of this study is to determine the nature of the hydrodynamics and the heat transfer characteristics as the phase angle between the geometry of the two plates is varied for a constant mean separation parameter. Quantitatively, the dimensionless pressure drop, ΔP, and Nusselt number, Nu, is found. An optimum configuration for which $\overline{Nu}/\Delta P$ is the largest is determined. The results are explained in light of the hydrodynamic and thermal characteristics indicated by the numerical simulation.

1   Introduction

Corrugated fin passages are used in a great variety of heat exchanger situations. The corrugations in the plates forming the channel have the function of stimulating the mixing of fluid in order to increase the heat transfer within the channel. Mixing reduces variations in the temperature in the working fluid, thereby steepening the temperature gradient near the boundaries and increasing the heat transfer between fluid and channel walls. In many applications, because of the small distance between plates or the viscosity of working fluids, the flow in this kind of channels may occur at low Reynolds numbers. Therefore, it is advantageous to develop laminar flows with good heat transfer characteristics for wavy channels.

Because of the wide application on plate heat exchangers, it is very important to find mechanisms of enhancement of heat transfer in channel flows. The easiest way to augment heat transfer is to increase the mass flow rate in order to enhance mixing by making the flow turbulent, but this is accompanied by a significant increase in the power required to move the fluid. Another suitable option is to adapt the channel geometry in order to promote mixing. This option has been studied by many investigators.

A number of investigators have developed studies on the performance of wavy channels: Blomerius and Mitra[1] obtained numerical solutions of the Navier–Stokes equations for the laminar and transitional flow in 2D and 3D wavy channels. They studied the geometric parameters for the best ratio of heat transfer to pressure drop for two-dimensional channels. Then they investigated for three dimensional channels of corrugated plates. Mehrabian and Poulter[2] looked for the effect of corrugation angle on the performance of a wavy channel when the plate spacing is fixed. Rush et al.[3] used visualization methods to study laminar and transitional flows. The experiments were developed in a channel with a 10:1 aspect ratio. Saniei and Dini[4] measured the local heat transfer and pressure drop from the corrugated portion of the bottom wall of a rectangular channel where the opposite top wall took one of the following three constructions: (1) plane; (2) covered with corrugations making an in-phase arrangement with the bottom wall; (3) covered with corrugations making an out-of-phase arrangement with the bottom wall. Sawyers et al.[5] used a combination of analytical and numerical techniques to study the effect of three-dimensional hydrodynamics on the enhancement of steady, laminar heat transfer in corrugated channels. For the case of two-dimensional sinusoidal corrugations with flow perpendicular to the corrugations, they found an increase in heat transfer as compared to the case of flat plates; there is increased advection near each stagnation point which, when combined with the asymmetry of the flow in the downstream direction, leads to a larger area-averaged heat transfer coefficient.

In this study we will consider modifications on the geometry of a two-dimensional channel. Variations of the phase angle between a pair of wavy plates will be considered in order to determine how this geometric parameter affects heat transfer both locally and as an average over one wavelength of corrugated plate.

2 Problem description

The problem to be studied is that of the flow in a channel comprised of two wavy fins, each one at a different isothermal temperature. Each plate forming

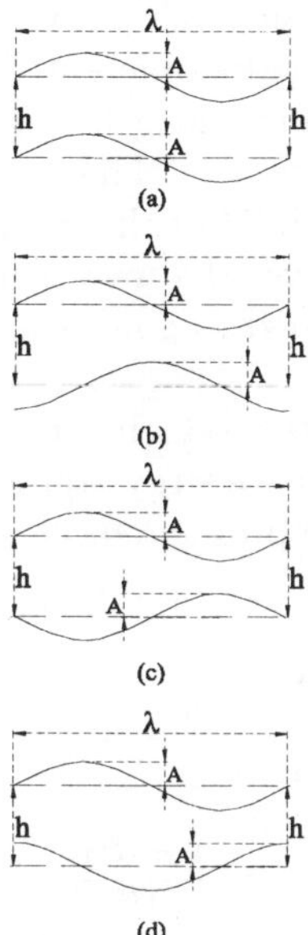

Figure 1. Wavy plates channel for four different phase angle between plates; (a) $\theta = 0$, (b) $\theta = \pi/2$, (c) $\theta = \pi$, (d) $\theta = 3\pi/2$.

the channel has sinusoidal shape and has the same amplitude and wave length as the other, but is allowed to have a phase difference with respect to the other. Figure 1 illustrates the geometric situation described above for four different phase angles: 0, $\pi/2$, π and $3\pi/2$. The geometric parameters involved are the wave length, λ, mean channel width, h, and sinusoidal wave amplitude, A.

The objective is to understand the hydrodynamics of the fully developed periodic flow and the corresponding heat transfer in the channel as a function of the phase angle between the plates. The flow under this situation is supposed to be two dimensional, laminar and steady state.

3 Numerical analysis

As explained in the previous section, the interest is to determine the flow and temperature fields over a cell of one wavelength of corrugated plates channel.

224

In order to avoid complications with inlet and outlet boundary conditions for such a cell (especially for phase angles different than 0 and π) a computational domain comprising a straight inflow length, several sinusoidal waves and an outflow straight section is considered. The velocity and temperature fields start with uniform values at the inlet of the domain and the temperature and velocity fields become periodic over the wave length after a few corrugations. The outlet straight section is added to guarantee free boundary conditions at the outflow.

The equations that govern the velocity, pressure and temperature fields are the continuity, momentum and energy equations for a Newtonian, steady state flow. In order to render the problem dimensionless, the following dimensionless parameters are chosen

$$\mathbf{x}^* = \frac{\mathbf{x}}{h} \tag{1}$$

$$\mathbf{u}^* = \frac{\mathbf{u}}{U} \tag{2}$$

$$p^* = \frac{p - P_0}{\rho U^2} \tag{3}$$

$$T^* = \frac{T - T_1}{T_2 - T_1} \tag{4}$$

where U is the mean velocity at the average channel width, h is the average channel width, T_1 is the lower plate temperature, T_2 is the upper plate temperature and P_0 is the constant pressure at the inlet. ρ represents the density of the fluid.

With this choice of dimensionless parameters the governing equations in dimensionless form, after dropping the asterisks, become

$$\nabla \cdot \mathbf{u} = 0 \tag{5}$$

$$(\mathbf{u} \cdot \nabla)\mathbf{u} = -\nabla p + \frac{1}{Re}\nabla^2 \mathbf{u} \tag{6}$$

$$(\mathbf{u} \cdot \nabla)T = \frac{1}{RePr}\nabla^2 T \tag{7}$$

where $Re = Uh/\nu$ is the channel width based Reynolds number, and $Pr = \nu\rho C_p/k$ is the fluid's Prandtl number. ν is the kinematic viscosity, C_p is the specific heat at constant pressure and k is the thermal conductivity of the fluid.

The dimensionless boundary conditions at the entrance of the channel are

$$u = \frac{h}{h - 2A} \tag{8}$$

$$v = 0 \tag{9}$$

$$T = \frac{1}{2} \tag{10}$$

$$P = P_0 \tag{11}$$

where A is the sinusoidal corrugation amplitude.

The boundary conditions on the lower plate surface are the nonslip and uniform temperature conditions

$$\mathbf{u} = 0 \tag{12}$$

$$T = 0 \tag{13}$$

The same kind of boundary conditions are established for the upper plate surface, but with a different value for the uniform plate temperature

$$\mathbf{u} = 0 \tag{14}$$

$$T = 1 \tag{15}$$

For the outflow section of the computational domain, we may allow for some extra transitional length to transform the wavy channel into a parallel plates channel. If we do so we can apply free boundary conditions at the outflow

$$\frac{\partial \mathbf{u}}{\partial x} = 0 \tag{16}$$

$$\frac{\partial T}{\partial x} = 0 \tag{17}$$

Additional to the dimensionless variables, the following dimensionless geometric parameters define the geometry.

$$\epsilon = \frac{h}{\lambda} \tag{18}$$

$$\beta = \frac{A}{h} \tag{19}$$

The ability to transfer heat is measured using the Nusselt number which is defined as

$$Nu = \frac{\partial T}{\partial n} \tag{20}$$

where T is the dimensionless temperature and n is the direction normal to the plate surface. The average Nusselt number is obtained by integrating the local Nusselt number over one wavelength and dividing this by the total area of one wavelength of plates.

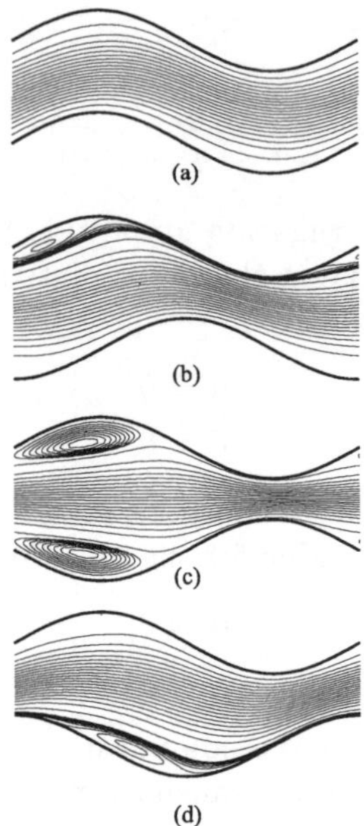

Figure 2. Streamlines in wavy channel flow for $Re = 100$, $Pr = 0.7$, $\epsilon = 0.3$, and $\beta = 0.3$. (a) $\theta = 0$, (b) $\theta = \pi/2$, (c) $\theta = \pi$, (d) $\theta = 3\pi/2$.

A general-purpose program for fluid mechanics and heat transfer, FIDAP, is used to solve the problem. The numerical technique is based on the finite element method which has the advantages of being flexible in its ability to adapt to complex geometries, permitting a distribution of grid density, and easy specification of boundary conditions on curved surfaces. A paved mesh of quadrilateral elements was used for meshing the computational domain. The typical mesh has 40 subdivisions in the y direction, 40 subdivision in the x direction over each wave period, 20 subdivisions in the x direction for the inlet and outlet straight sections. Several grids were tested for grid independence.

4 Results

The results obtained in this paper were compared with those of Sawyers et $al.$[5] and very good agreement was found for similar geometries and Reynolds numbers.

From the numerical results obtained, it was observed that the flow and

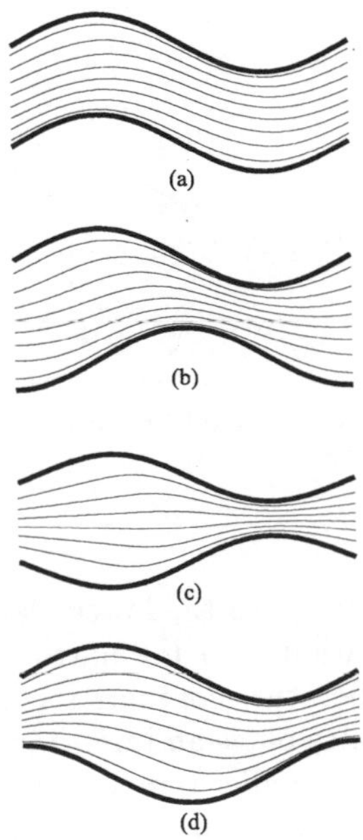

Figure 3. Temperature contours in wavy channel flow for $Re = 100$, $Pr = 0.7$, $\epsilon = 0.3$, and $\beta = 0.3$. (a) $\theta = 0$, (b) $\theta = \pi/2$, (c) $\theta = \pi$, (d) $\theta = 3\pi/2$.

temperature fields became periodic after just one corrugation. The results to be presented now correspond to velocity and temperature fields of the third corrugation provided that the conditions are those of a periodic flow.

Figure 2 shows the streamlines for four different phase angles. For $\theta = 0$ the fluid flows without recirculation occurring. However, it can be mentioned that the streamlines tend to separate from the wall in the sections of the channel where the fluid has just passed an obstacle and to get nearer to the walls in sections of the channel where the fluid moves toward an obstacle. For $\theta = \pi/2$, there is no boundary layer separation over the lower plate but the motion of the streamlines closer or farther from the lower wall is visible. Things are very different in the upper plate where there is separation of the boundary layer and creation of a recirculation bubble after the narrower part of the channel, the boundary layer reattaches before entering a new strait. For $\theta = \pi$ the flow becomes symmetric with respect to y. There are separated boundary layers after the narrow section of the channel and those boundary layers reattach before entrance to a new strait. The case $\theta = 3\pi/2$ is very

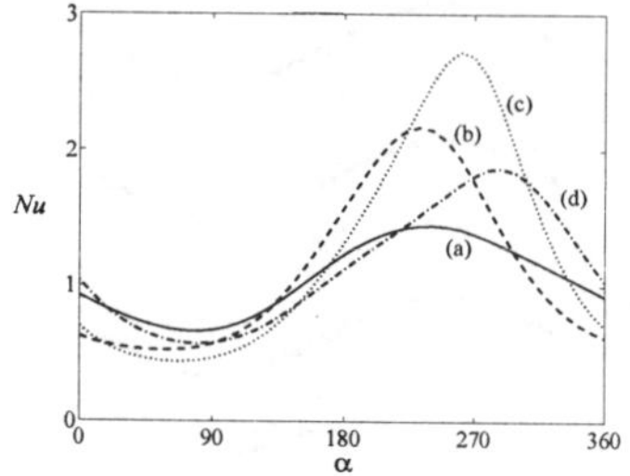

Figure 4. Nu as a function of α for $Re = 100$, $Pr = 0.7$, $\epsilon = 0.3$, and $\beta = 0.3$. (a) $\theta = 0$, (b) $\theta = \pi/2$, (c) $\theta = \pi$, (d) $\theta = 3\pi/2$.

similar to the case $\theta = \pi/2$ but with the lower plate experiencing a separated boundary layer and recirculation after the narrower section of the channel.

Figure 3 shows the temperature contours for the same four values of θ. It is seen that the hydrodynamics plays an important role on the determination of the temperature field with smaller temperature gradients near the wall in recirculation regions. This temperature distribution makes the local Nusselt number small in regions of fluid recirculation. Figure 4 shows the values of the local Nusselt number over the upper plate as a function of sine wave angle, α, for various values of dephasing. For $\theta = 0$ the ratio between the maximum and minimum Nu values is smaller since the flow does not present recirculation zones where the Nusselt number is very small. This is also true because the plate transversal section is constant. For the case $\theta = \pi/2$, the ratio between the maximum and minimum local Nusselt number increases because the appearance of a recirculation zone and the appearance of a strait section of the channel. In the case $\theta = \pi$ the ratio of maximum to minimum Nusselt numbers is the highest, in fact, the lowest local Nusselt number occurs in this case, since the recirculation bubble is largest, and the highest Nusselt number occurs also in this case, since the strait is the narrowest. For case $\theta = 3\pi/2$, the ratio of maximum to minimum local Nusselt numbers is reduced since there is no recirculation zone over the upper plate. The highest local Nusselt number occurs in the narrowest part of the channel.

Figure 5 shows the variation of the average Nusselt number over one wavelength as a function of dephasing angle, θ. The maximum overall heat transfer coefficient occurs for the case $\theta = \pi$. The result illustrates that the dominant effect is the narrowing of the channel. The cases $\theta = \pi/2$ and $\theta = 3\pi/2$, as expected, give the same overall Nusselt number. The case $\theta = 0$ gives the lowest average Nusselt number since the channel cross section is

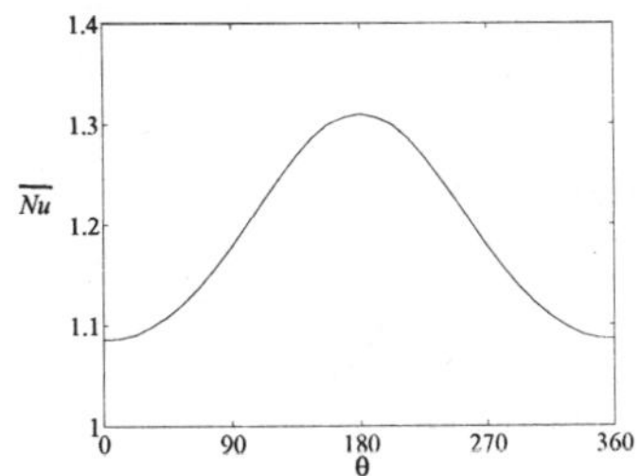

Figure 5. $\overline{Nu}$ as a function of θ for $Re = 100$, $Pr = 0.7$, $\epsilon = 0.3$, and $\beta = 0.3$.

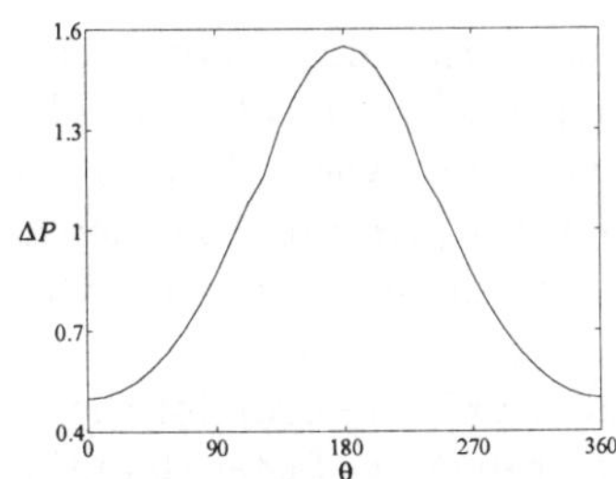

Figure 6. Pressure drop in one wave period as a function of θ for $Re = 100$, $Pr = 0.7$, $\epsilon = 0.3$, and $\beta = 0.3$.

constant.

Figure 6 shows the variation of the pressure drop over one wavelength as a function of dephasing angle. The maximum pressure drop occurs in the case $\theta = \pi$. This is caused by the acceleration of the fluid occurring in the strait. The case $\theta = 0$ gives the lowest pressure drop since the mean fluid velocity is constant over the whole channel wavelength.

In the case of the ratio of average Nusselt number to pressure drop over one wavelength, shown in Fig. 7, we observe that the maximum occurs in the case $\theta = 0$. This reflects the fact that the growth of ΔP as we move toward $\theta = \pi$ is much more pronounced to that of $\overline{Nu}$ since the pressure drop grows to the square of the velocity and is also due to the fact that the growth of the recirculation zone around $\theta = \pi$ limits the growth of $\overline{Nu}$.

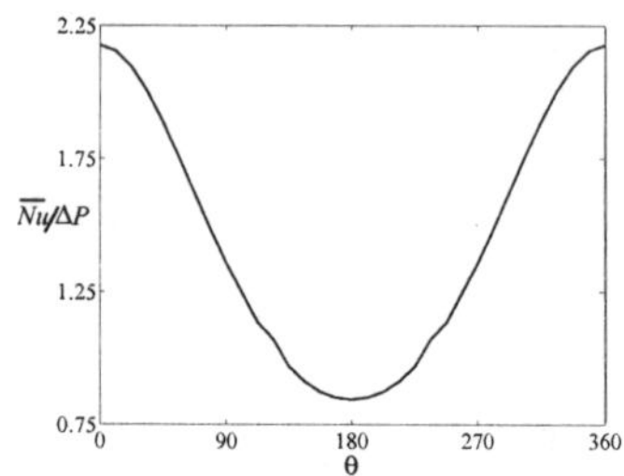

Figure 7. $\overline{Nu}/\Delta P$ as a function of θ for $Re = 100$, $Pr = 0.7$, $\epsilon = 0.3$, and $\beta = 0.3$.

5 Conclusions

The results of the previous section have shown the increase of heat transfer that occurs in a channel formed by pair of sinusoidally corrugated plates. As compared to the case of flat plates, where the average Nusselt number is unity, all cases of corrugated channel presented an increase in the overall Nusselt number. Of all the corrugated channels, the case where the dephasing between plates is 180 degrees $(\theta = \pi)$ gave the largest average Nusselt number but also gave he largest dimensionless pressure drop for the same Reynolds number. It happens that the pressure drop grows faster than the average Nusselt number as we approach $\theta = \pi$ because there is also a larger recirculation zone as we approach $\theta = \pi$. As a result of this, the channel configuration that gives the largest ratio average Nusselt number to pressure drop is the case $\theta = 0$.

Acknowledgement

The authors wish to acknowledge the support of the Mexican Council for Science and Technology (CONACyT).

References

1. H. Blomerius and N. K. Mitra, *Numerical Heat Transfer, Part A* **37**, 37 (2000).
2. M. A. Mehrabian and R. Poulter, *Applied Mathematical Modelling* **24**, 343 (2000).
3. T. A. Rush, T. A. Newell and A. M. Jacobi, *International Journal of Heat and Mass Transfer* **42**, 1541 (1999).
4. N. Saniei and S. Dini, *Heat Transfer Engineering* **14**, 19 (1993).
5. D. Sawyers, M. Sen and H.-C. Chang, *International Journal of Heat and Mass Transfer* **41**, 3559 (1998).

ANALYSIS OF THE NUMERICAL SIMULATION OF THE FILLING SYSTEM USED IN PRODUCTION OF ALUMINUM PISTONS

A. R. CABRERA-RUIZ, A. SORIA-VIÑAS
Centro de Desarrollo Tecnológico, Pistones MORESA, S. A. de C. V.,
Carr. Panamericana Km. 284, 38110 Celaya, Gto., México
E-mail: asoria@uniko.com.mx, acabrera@uniko.com.mx

A. HERNÁNDEZ-GUERRERO
Facultad de Ingeniería Mecánica Eléctrica y Electrónica, Prolongación Tampico S/N,
Apartado Postal 215-A, 36730 Salamanca, Gto. México
E-mail:abelh@salamanca.ugto.mx

M. CAUDILLO-RAMÍREZ
Instituto Tecnológico de Celaya, Av. Tecnológico y A. García Cubas S/N,
Apartado Postal 57, 38010 Celaya, Gto. México
E-mail:martinc@itc.mx

Permanent-molding casting of aluminum-based pistons is affected by defects that cause rejection in quality-check control. Current scientific and technical literature plus *in-situ* experiences have allowed the identification of the causes of such problems. One of these causes is a direct consequence of a poor feeding system, which produces excessive velocities and critical conditions that generate turbulence and detachment of the thin film oxide lining of the feeding system, as well as bubble trapping. In this work, by means of numerical simulation, different feeding systems currently in use were studied (the analysis does not account for bubble trapping.) Modifications are suggested based on the analysis of these results. These modifications have a direct impact in the geometry of the feeding system, avoiding stagnation points as well as excessive velocities, giving as a result important reductions in the rejection of defective pistons.

1 Introduction

The manufacturing of pistons involves the process of aluminum melting, in which the pieces that are the basic geometry of the piston are produced. To produce a piston according to the design specification it is necessary that the final product of the melting process shows no defects in the initial inspection, such as porosities, inclusions, contractions and marks due to mishandling of the material, amongst others. Other defects will be produced afterwards and will show after the final stage of machining the piston, such as segregation, inclusion of material foreign to the alloy, macro and microporosity, etc. Due to these defects, the criteria that control the allowed porosity are very rigid, since they will directly affect the mechanical properties of the final product, as well as its surface finish. To fulfill the quality and design requirements that are demanded by the world engine manufacturers, it is necessary to have good control of the process, as well as to avoid defects in the raw material.

The melting process is complex due to the large numbers of variables involved which make it difficult to control. Moreover, the high costs involved in instrumenting a mold to monitor the process *in-situ* make it necessary to search fo tools that could provide information about the behavior of the metal inside the mold as well as its possible optimization[1]. One of these tools is numerical simulation which allows us to determine how the process occurs and analyze the origin of the defects.

In our industry this need has surfaced as an answer to the many problems that have been detected in the melting process, impacting the rejection of cast pieces. The porosity in the cast billets is one of the defects that most affect rejection, in comparison to other common defects. Its origin has been blamed on the hydrogen concentration in the Al–Si alloy[2-4]. It is true that high levels of hydrogen concentration induce porosity during solidification, however, the numerical simulation has helped to detect the formation of porosity at the beginning of the metal feeding. In the technical literature the same has been reported[5-7].

This work presents the results of the simulation of the feeding process to form the basic geometry of the piston. These results have helped to detect and correct steps in the feeding process, which has resulted in decreasing the rejection of cast billets.

2   The feeding process

The casting manufacturing process is carried out in a melting cell that has a turning device with 3 induction ovens that continuously provide metal raw material to a couple of large bowls that are at the end of a robotic arm; these bowls then feed the four molds of the two molding machines (see Fig. 1).

The metal that has left the ladle enters the mold by means of a feeder known as the sprue. There are different types of these feeders, depending on the application and size of the piece to cast. In the case of the pistons, considered small pieces, the known types are very similar, as shown in Fig. 2. For the design of the feeders several authors propose strategies to get good results[6-9], however the cases on which they base their suggestions are too general and for relatively large pieces, as compared to pistons. Therefore in this analysis their pragmatic suggestions are taken into account but looking more closely to the particular case of a small piece.

As the feeding process is controlled by the robotic arm, it is expected to have repeability; however, it has been noted that other problems arise that bypass this control. Therefore, for the analysis, it is necessary to check all parts in the process: the feeding bowl, the mold, as well as the metal material fed to the mold.

The analysis starts by reviewing the behavior of the aluminum flow that is provided by the bowl in response to the movement of the robotic arm. The response of the flow of liquid metal is due to the position of the robotic arm, as well as to the time required to fill the mold. Figure 3 shows the filled volume as a function of the angle of reference. This information is quite important since it provides the initial conditions for the modeling of the metal feeding to the mold.

Figure 1. Melting cell.

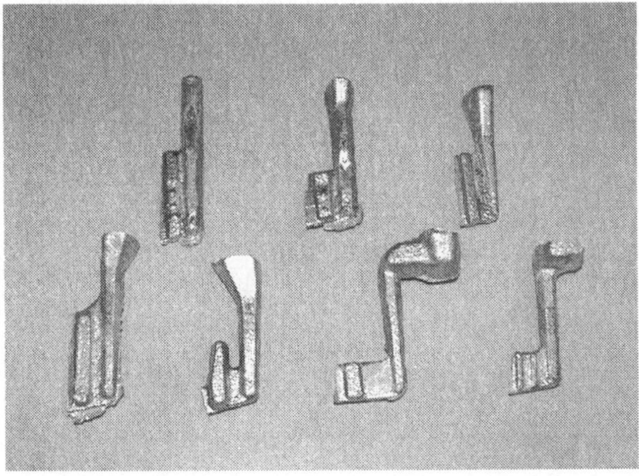

Figure 2. Sprue types for piston casting.

The feeding initial velocity is calculated to approximate the conditions of the metal at the entrance of the mold. It is quite evident that the geometry of the mold plays a very important role in the behavior of the flow of metal, which is even more remarkable once the results of the modeling are obtained. In Fig. 4 a scheme of the feeder model under analysis is presented, as well as its constitutive parts.

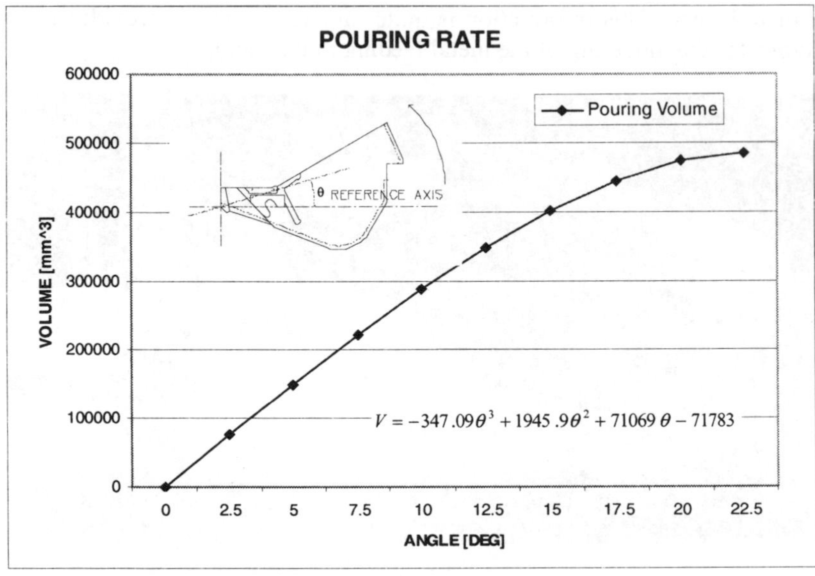

$$V = -347.09\theta^3 + 1945.9\theta^2 + 71069\,\theta - 71783$$

Figure 3. Ladle used to feed the metal to the permanent molds.

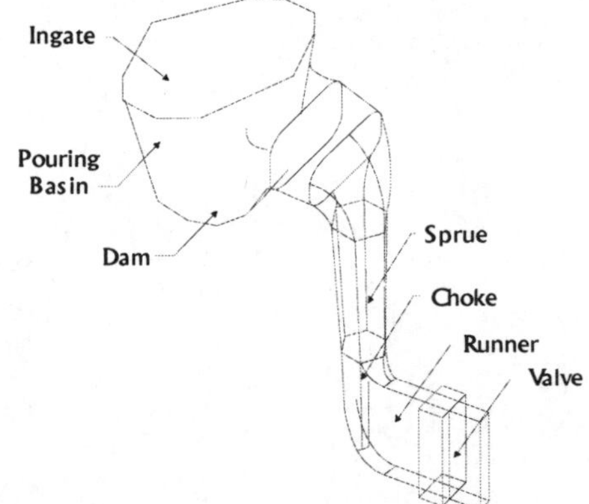

Figure 4. Parts of the feeding system.

3   Equations for the feeding model

To analyze the feeder, as well as the complete cast billet, it is required to establish a model that considers the phenomena that govern the behavior of the liquid metal that flows throughout the piece. What follows is the mathematical model on which the code for the analysis is based.

To simulate the flow of the liquid metal a generalized mathematical model for a control volume is used. It is assumed that the metal behaves as a Newtonian fluid[10] during the feeding process up to the moment the filling of the mold is finished, that is, the metal will still be in the same phase up to then.

The mass balance that governs the process is given by:

$$\frac{\partial \rho}{\partial t} + \nabla \cdot \rho \bar{v} = 0$$

By using the stress–strain relations, considering forces due to shear and normal stresses and constant properties, the governing momentum equation is:

$$\rho \frac{D\bar{u}}{Dt} = \rho \bar{F} - \nabla p + \mu \nabla^2 \bar{v}$$

For the energy balance the governing equation is:

$$\frac{DT}{Dt} = \alpha \nabla^2 T + \Phi$$

where the last term is the so-called "dissipation function", which is considerable in a problem like this. These main governing equations are solved by means of finite differences via the MAGMAsoft code[11].

The transient complete model uses around 1,600,000 cells (control volumes). Normally geometries that can be considered symmetrical are cut off from analysis, since half the mesh could be considered superfluous and if high timesteps are used, they will only add to produce overflow problems; however, due to the very thin geometry of some of the zones of the feeder it is very critical to determine high gradients where they occur. Therefore the mesh was done over the whole unit.

In areas not considered critical for this study, such as the piston itself, as well as other components that are part of the unit (such as the raisers), a coarse mesh was used. For areas considered very critical, such as the choke, runner and valve, the mesh was decreased until it could be assured that the contact between control volumes occurred not only at one or two points, but that it was fully of the face-to-face type (care was taken so that the areas of the control volumes were interconnected.) Therefore, to avoid convergence problems the code allows approximating the cells to cubic geometry. This is done as much as the geometry of the pistons allows. The first part of the work was to extensively optimize the mesh to be used in the critical zones, until no appreciable variation in the results was found. Once the mesh was set for the critical zones, normally no problems were found when a geometry change was implemented in other zones of the feeder to determine if the change produced a strong effect in the results (for instance when the angle of falling was varied). The model does not consider turbulent regimes. The reason to use this code is the ease it offers to solve systems with serious convergence problems, as compared to using finite element.

4 Results

The numerical results showed a vortex being generated at the beginning of the filling process, as well as splattering (see Fig. 5). Both things produce formation and excessive breaking of the thin aluminum oxide films that form on the surface of the metal that is exposed to the ambient, having a negative impact on the final quality of the aluminum alloy properties.

In practical terms, air trapped in the feeder causes air bubble formation; the bubbles that cannot reach the surface generate macro porosity in the piston, causing internal formation of oxides, as well. The code does not specifically account for bubble formation, however, speculatively, it can be seen that the flow distribution throughout the feeder system shows spaces with no fluid metal that eventually will be filled; these spaces must have air or gases. It is observed that in the beginning of the filling process in the lower part of the sprue not all the volume is occupied by the falling melted metal. It is at this moment where air and other gases are trapped in this region. Eventually those gases are dragged by the filling liquid and taken into the piston being formed. See Figs. 5 and 6.

Trying to avoid contraction while the metal starts to solidify, the entry zone to the main part of the piece to cast suffers a severe area reduction. This reduction causes an excessive increase in speed of the flow of metal in this zone, provoking an impact of the flowing metal with the mold exactly at the point where the flow has to change directions and split. These velocities cause rupturing of the ceramic paint (vermiculite and graphite) that is used to avoid thermal shock between the metals, as well as to protect the steel mold. It is found that velocities over 75 cm/s tend to tear excessively the mold and in some cases completely break the protective paint, which in turn produces inclusions that most likely result in rejections of the final pieces (see Fig. 7).

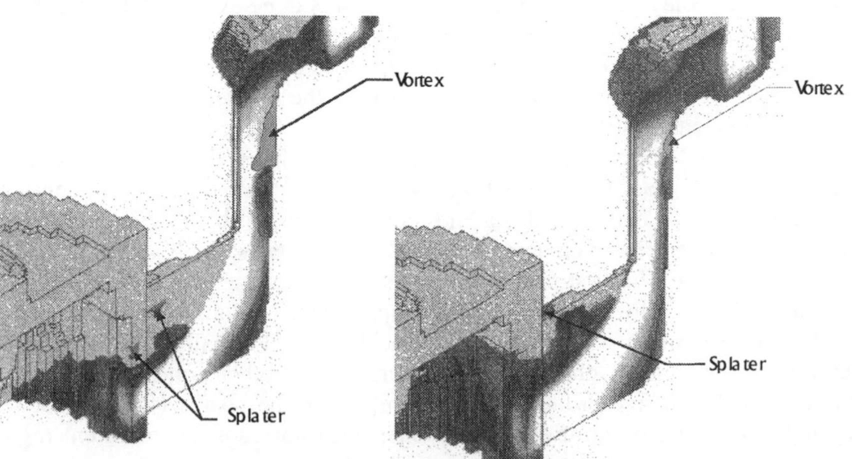

Figure 5. Vortex generation in the sprue.

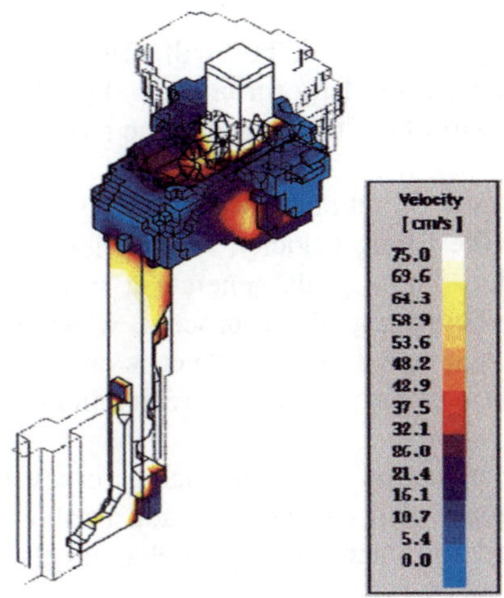

Figure 6. Velocity distribution in the initial stage of the filling process.

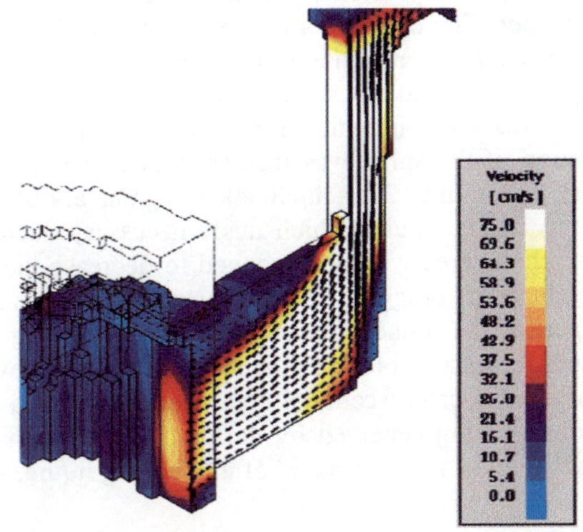

Figure 7. Excessive velocities at the entrance zone to the mold.

The behavior shown in the previous figures were observed during the simulation runs in which different geometric changes were entered, as well as changes in the radii as a result of increases in the falling angle of the sprue. Other changes implemented, such as the distribution and size of the feeder valve, and a thinner pouring basin, helped to make the problems of vortex generation and excessive splattering disappear. However, the entry velocity could not be changed drastically, since the ladle velocity cannot be easily changed (as it would affect the production capacity). To simulate more complex geometries could decrease the velocity of the filling flow metal, but will also generate counter-current flows, increasing the problems caused by turbulence. This leads us to conclude that the changes in the lower part of the feeder cannot be too severe.

Due to the constraint of not being able to make severe geometry changes and to the need to decrease the filling velocity, small grids have been implemented at the entrance of the feeder. These grids generate a significant decrease in the speed, specially at the very beginning of the process, where the root of the problems is. The use of these grids, coupled to the geometric changes has generated a surprising decrease in the rejection of pieces due to porosity and other defects.

Since this is an industrial problem, there are several things on the technical solutions that were made that can not released explicitly. However the following comments can be made in regards to global changes that were implemented thanks to this work.

To follow the changes implemented, please refer to Fig. 4 for the technical names used in the feeder. The entrance area (ingate) to the feeder was reduced, as well as the pouring basin. The connecting section between the pouring basin and the sprue was changed so as to make a softer fall. The radii in these sections were slowly modified and fed back into the numerical simulation to determine their effects. The inclination of the sprue was increased as much as possible given the space and geometry constraints. This inclination, again, allowed a softer fall to avoid splashes of the falling liquid (which made necessary decreasing the runner section length). The choke area was also changed to decrease the bottleneck effect and decrease the chances of dragging ceramic protective paint. The height of the valve (which is a part of the runner section) was enlarged. Originally, the idea of having this valve was to act as a break to the falling fluid. However, by increasing its height and enlarging its cross sectional area it acts now as a gas trapper. One final comment in changes implemented by the modeling was that a metallic grid was added to the ingate to act as a "breaker" of the feeding liquid, slowing it.

5   Conclusion

The code simulation offers surprising results in processes in which it is very complex to make *in-situ* measurements and monitoring, such as in the manufacturing of aluminum pistons. The results of this work provide information to make changes that will render a more stable production, with less rejection problems due to bad mold design or conditions under which the filling process is carried out.

The modification of the filling systems as a result of the simulation and the feedback to optimize the geometries chosen has allowed visualization of the final outcome. This in turn allowed arriving at decisions that have lowered the rejection percentage and the waste of material

The reduction in the waste percentage has been from an initial value on the order of 25% to around 7%, due to the decrease in velocities of the metal, as well as modifications in the geometry of the feeders. However, at this point a proposal is needed on how to reuse the leftovers of aluminum, since there is a contamination of the leftover metal by the small metallic grid that has been implemented.

Acknowledgements

The authors would like to thank the great participation of the engineering team of the Centro de Desarrollo Tecnológico de Pistones MORESA S. A. de C. V., as well as the never ending support to use equipment and technical literature resources at the center. The patience and time given by Miss Linda Escobar is also greatly appreciated.

References

1. David C. Schmidt, *Casting Solidification Control and Mold Design Using Computer Simulation* (American Foundrymen's Society, IL, USA, 1995).
2. J. Huang, T. Mori, J. G. Conley, "Simulation of Microporosity Formation in Modified and Unmodified A356 Alloy Casting", 4th. International Conference on Permanent Mold Casting of Aluminum, Nov, 1997.
3. J. Huang and G. J. Conley, *Computer Simulation of Pore Size and Shape for Equiaxed Aluminum Alloy Casting* (Northwestern University, 1998).
4. W. S. Hwang, "Modeling of Fluid Flow", *Computer Applications in Metal Casting* (Pittsburgh University, USA, 1996).
5. V. Desa Prateen, *Modeling of Combined Fluid Flow and Heat/Mass Transfer* (Georgia Institute of Technology, 1983).
6. John Campbell, *Castings*, Chap.2–4 (Butterworth-Heinemann, Birmingham, UK, 1991).
7. Philip Sanford, "Optimization of Gating Systems", 4th International Conference on Permanent Mold Casting of Aluminum, Nov., 1997.
8. John Campbell, "The Rules for Good Casting", *Modern Casting*, April, 1997.
9. Jeff Meredith, "Fundamentals of Gating and Feeding Nonferrous Alloys", *Modern Casting*, February, 1999.
10. J. Szekely, *Fenómeno de Flujo de Fluidos en Procesamiento de Metales*, Chap. 2–3 (Limusa, México, 1998).
11. MAGMAsoft version 2.1, *Physical Background Manual*, Chap. 8, 1995.

ISOTHERMS IN A SOLID WITH A TILTED FRACTURE AND THEIR EFFECTS ON LOCAL FLUID MOTION

A. MEDINA, E. LUNA, C. PEREZ-ROSALES

Instituto Mexicano del Petróleo, Eje Central Lázaro Cárdenas 152,
Apdo. Postal 14-805, 07730 México, D.F., México

C. TREVINO

Departamento de Física, Facultad de Ciencias, UNAM,
Apdo. Postal 70-564, 04510 México, D.F., México

In this work we analyze the quasisteady state problem of heat transfer in an impermeable solid by considering a background, constant, vertical temperature gradient, G, and the existence of a tilted slot either solid-filled or fluid-filled and located at its middle part. We study cases when a solid medium in the fracture has smaller thermal conductivity than the solid matrix and when a fluid medium has a thermal conductivity smaller than the surrounding solid. In this latter case, which is the most important from the practical point of view such as in the study of oil reservoirs, the heat diffusion in the solid is almost independent of the fluid part and therefore the boundary conditions on the fracture walls are obtained only as a solution of the heat transfer problem along the solid matrix. The energy heat equation is solved numerically for each part of the solid (besides the fracture) in a two-dimensional domain, by employing a paralellized finite differences code in order to find the temperature profiles (isotherms) around the fracture. Finally, these results are also employed in the study of the thermal convection in the fluid when the fracture is very long compared with the aperture size. We have obtained analytical solutions for this particular case.

1   Introduction

Many oil and water reservoirs are naturally fractured, i.e., the near impermeable rocks where the fluids are contained are fissured. Moreover, these reservoirs are under the action of a vertical geothermal gradient which is approximately constant. In a typical reservoir, the fluids in the tilted fractures are nearly isolated from the porous matrix. Thus, the problem of thermal convection in each fracture is very interesting and, consequently, it is very important in order to understand the convection in the reservoir and related phenomena such as the dispersion of several types of materials and diagenesis[1,2].

Some efforts have been made to study the fluid behavior inside a fluid-filled, tilted fracture under a constant vertical gradient in a nearly impermeable reservoir which can be considered as a solid matrix[1,3]. The direct approximation in these works, employed to study the quiescent and convecting states, was the simple projection of the vertical temperature gradient on the tilted walls of the fracture. However, this approximation can be seen as local and is not correct when the fracture length is small or when the fluid motion in the whole fracture zone is of interest.

In this work we analyze the steady problem of the fluid motion in a two dimensional tilted fracture, located at the middle part of a vertical solid, by considering a background vertical temperature gradient, G, at the edges of the solid and under the acceleration of gravity, g. We study cases when a solid in the fracture has smaller thermal conductivity than the solid matrix and when the fluid has thermal conductivity smaller than the solid. In this case, which is the most useful in the study of oil reservoirs, heat diffusion in the solid is nearly independent of the fluid part, and therefore the boundary conditions on the fracture walls needed to solve the fluid equations are obtained as a solution of this problem. Finally, these results are also employed in the study of the thermal convection in the fluid at low Rayleigh numbers. The structure of this work is as follows. In the next section we formulate the problem and the corresponding governing equations. There we also show the use of a paralellized finite differences code in order to find the temperature profiles (isotherms) around the fracture and inside it. These results are used in section 3 to study the convective fluid motion. Finally, in section 4 we give the conclusions and some comments related to future work.

2 The problem

A schematic representation of the fracture in the solid matrix is shown in Fig. 1 where the dimensions of the solid matrix, of the fracture and the coordinate systems inside and outside the fracture are also shown. Briefly, we have considered a fracture with an angle of tilt ϕ, aperture size d, infinite width and finite length $L = H/\sin\phi$; the dimensions of the solid matrix are length D and height H. The simplest form to set a background thermal gradient is, as shown in Figure (1), by setting a temperature T_u at the top and a temperature T_l at the bottom and therefore, $\Delta T = T_u - T_l > 0$. The temperature gradient at the lateral walls of the large enough solid matrix is $G = \Delta T/H$. The fluid has thermal expansion coefficient β, kinematic viscosity ν, density ρ, specific heat c, thermal conductivity of the solid k_s, thermal conductivity of the fluid k_f, thermal diffusivity of the fluid $\alpha_f = k_f/\rho c$ and finally, the ratio of the thermal conductivities is $\varepsilon = k_f/k_s$, assumed to be very small compared with unity.

The isotherms in the solid will be parallel to the horizontal far away from the fracture but close to it their profiles will be deformed in accordance with the boundary conditions corresponding to the continuity of temperature and

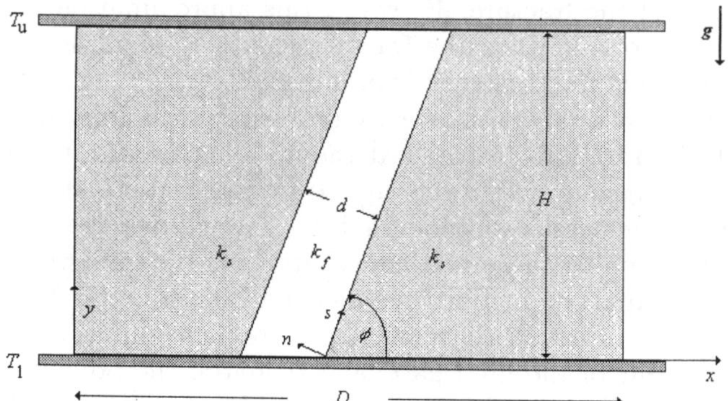

Figure 1. Schematic representation of a tilted fracture in a solid of length D, height H and infinite width. The angle of tilt of the fracture is ϕ, the aperture size d and length $L = H/\sin\phi$.

heat fluxes. Thus, any solution for the flow is strongly dependent of the boundary conditions (temperature profiles along each fracture wall). In order to know the temperature profiles, we assume that the thermal conductivity of the fluid is very small compared with that of the solid ($\varepsilon \to 0$) and therefore the fluid motion in the fracture does not affect the temperature distribution outside the fracture. In this case, it is sufficient to solve the heat conduction problem in the solid and apply the resulting temperature distributions to both sides of the fracture.

The procedure to solve the heat diffusion problem in the solid separates it into two regions: the side to the right of the fissure, (x_1, y_1), and the side to the left, (x_2, y_2). In Fig. 2 we show schematically the regions under consideration. The nondimensional energy equations in each region of the solid matrix are:

$$\frac{\partial^2 \theta_i}{\partial x_i^2} + \frac{\partial^2 \theta_i}{\partial y_i^2} = 0 \quad \text{for } i = 1, 2, \tag{1}$$

where $\theta_i = (T_i - T_l)/(T_u - T_l)$ and all coordinates have been made dimensionless with the height H: $x_i \to x_i/H$ and $y_i \to y_i/H$. The corresponding nondimensional boundary conditions are:

$$\theta_1\,(y_1 = 0) = \theta_2\,(y_2 = 1) = 0, \tag{2}$$

$$\theta_1\,(y_1 = 1) = \theta_2\,(y_2 = 0) = 1, \tag{3}$$

$$\theta_1 \left(x_1 = 0 \right) = 1 - y_1, \ \theta_2 \left(x_2 = 0 \right) = y_2 \tag{4}$$

$$\left. \frac{\partial \theta_1}{\partial n_1} \right|_{\phi_1} = \varepsilon \left. \frac{\partial \theta_f}{\partial n_1} \right|_{\phi_1}, \ \left. \frac{\partial \theta_2}{\partial n_2} \right|_{\phi_2} = \varepsilon \left. \frac{\partial \theta_f}{\partial n_2} \right|_{\phi_2}, \tag{5}$$

where n_i are the nondimensional normal coordinates to the fracture wall for region i and θ_f represents the nondimensional temperature in the fluid inside the fracture. ϕ_i indicates that the gradients are considered along the fracture facing region i.

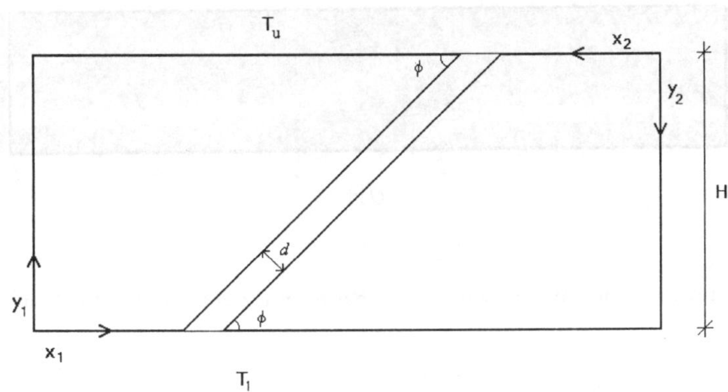

Figure 2. Coordinate axes for each region of the solid matrix. The right hand side is referred to the coordinate system x_2, y_2 and the left hand side is referred to the coordinate system x_1, y_1.

2.1 Numerical method

To solve the energy equation in each solid part we have used a numerical method based on finite differences. The order of accuracy in the method is $O(\Delta x)^2 \sim O(1/40)^2$. Parallelization was applied by using a pseudotransient form of the energy equations (1) for the solid with a cubic spline interpolation technique.

Figure 3 shows a plot of the isotherms for the case when the solid has much larger thermal conductivity than the fluid, $\varepsilon \to 0$, with an angle of tilt $\phi = 45°$. In this limit ($\varepsilon \to 0$), the energy equations in the solids are decoupled from the energy equation for the fluid inside the fracture. For a finite value of ε, both phases have to be solved simultaneously. The nondimensional temperature profiles along the fracture can be obtained, as shown in Fig. 4 as a function of the nondimensional longitudinal coordinate, $\xi = s/L$, along the walls of the fracture. In this case we note the temperature distribution along the fracture is nonlinear and strongly dependent on the whole geometry. The numerical treatment employed in this work assumes that $D >> H \cot(\phi)$, which means that the isotherms in the solid are almost horizontal far away of the fracture. For $\varepsilon \to 0$, the temperature gradient in the fluid is very different than that of the solids. Strong temperature gradients of order $\Delta T/d$ arise, producing important convective motions in the fluid, studied briefly in the next section.

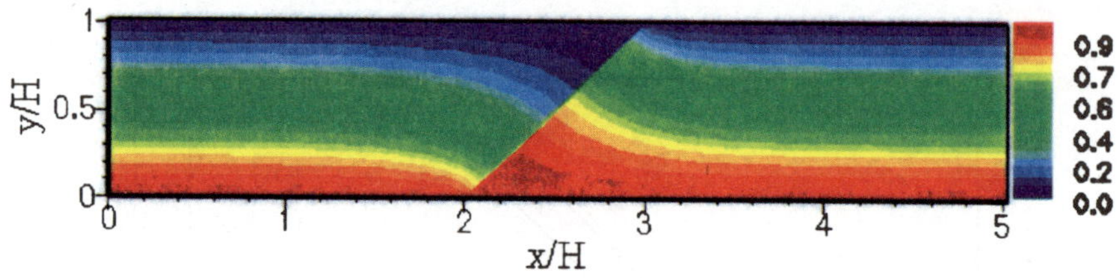

Figure 3. Schematic view of the isotherms in the solid and in the fissure.

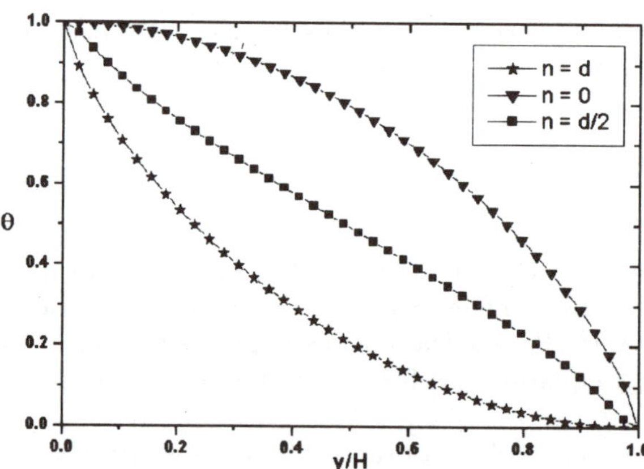

Figure 4. Dimensionless temperature as a function of the coordinate y/H on the fracture walls and in the fluid part.

3 Local solution for the fluid

By considering the solution obtained from the problem of the heat transfer along the tilted plates, we note that locally there is a zone (around $\xi = 1/2$) where the temperature gradients are almost constants and the flow is almost parallel to the fracture. In this case, the values of the temperature along the fracture's walls were well determined from the complete problem of heat conduction in the solid and in the fluid, as previously shown.

Assuming the flow to be parallel to the fracture walls, we can derive a simple theory to obtain the velocity profiles in the fluid. The momentum equations for the fluid are, assuming the coordinate system (n, s) as the directions normal to the fracture and along this one, respectively,

$$0 = \rho g \sin \phi + \mu \frac{\partial^2 u}{\partial n^2} - \frac{\partial p}{\partial s}, \tag{6}$$

$$0 = \rho g \cos \phi - \frac{\partial p}{\partial n}, \tag{7}$$

and the energy equation is

$$\rho u c \frac{\partial T}{\partial s} = k_f \left(\frac{\partial^2 T}{\partial s^2} + \frac{\partial^2 T}{\partial n^2} \right). \tag{8}$$

The boundary conditions are

$$u(0) = u(d) = 0, \tag{9}$$

$$T(0) = T_2 + G_2 s, \tag{10}$$

$$T(d) = T_1 + G_1 s, \tag{11}$$

where G_1 and G_2 are given by the numerical method. The overall mass conservation is given by

$$\int_0^d u\,dn = 0. \tag{12}$$

Using the Boussinesq approximation (which is the substitution of the density $\rho(T)$, by the value $\rho(T) = \rho_0[1 - \beta(T - T_0)]$ in buoyancy terms in the momentum equations) and the dimensionless quantities $(\eta, \xi) = (n/d, s/L)$, $\theta = (T - T_0)/\Delta T$, $p^* = p/(\rho_0 dg \cos(\phi))$, with $p^* = \eta + \beta \Delta T \tilde{p} + \tan(\phi)\xi/\Gamma$, $u^* = u\Gamma \cos(\phi)/(g\beta\Delta T d^2 \sin^2(\phi)/\nu)$, $\Gamma = d/L$ and the Rayleigh number $Ra = g\beta\Delta T d^3 \sin(\phi)/(\nu\alpha_f)$, the nondimensional equations reduce to

$$\frac{\partial^3 u^*}{\partial \eta^3} - \frac{\partial \theta}{\partial \eta} = -\frac{\Gamma}{\tan(\phi)} \frac{\partial \theta}{\partial \xi} \tag{13}$$

$$\frac{\partial^2 \theta}{\partial \eta^2} = -\Gamma^2 \frac{\partial^2 \theta}{\partial \xi^2} + Ra \tan(\phi) u^* \frac{\partial \theta}{\partial \xi}. \tag{14}$$

Here we have eliminated the nondimensional pressure term $\tilde{p}$. In the limit $Ra \to 0$ and $\Gamma \to 0$, the temperature distribution in the fluid is linear, $\theta = -C_1(\xi)\eta + C_2(\xi)$, with C_1 and C_2 of order unity. In this limit, the solution to Eq. (13) is given by

$$u^* = C_1(\xi) \left\{ \frac{\eta}{12} + \frac{\eta^2}{4} - \frac{\eta^3}{6} \right\}. \tag{15}$$

This result shows that by knowing the temperature profiles and gradients on the walls of the fracture, there is always a convective flow inside the fracture, regardless of how small the Rayleigh number is.

4 Conclusions and future work

In this work we have analyzed the steady problem of the heat transfer in a fluid inside a tilted fracture imbedded in a solid under a vertical temperature gradient, with the thermal conductivity much larger than that of the fluid, as in a typical oil reservoir. We have found that the isotherms are not only the projection of the temperature gradient on the walls, but instead their structure is very complex and depends on the aspect ratio of the fracture, the ratio of the thermal conductivity of the fluid related to that of the solid and the inclination angle. A simple theory using a parallel flow inside the fracture gives the velocity and temperature profiles inside the fracture, which has to be tested with experiments and detailed numerical calculations using the full Navier–Stokes equations. Another problem to be considered is when instead of a fluid-filled fracture we have a tilted porous layer saturated of a fluid in a solid of low permeability matrix[4]. Efforts in this direction are just now in progress.

Acknowledgment

This work has been supported by the Grant FIES 98 58-I of the Instituto Mexicano del Petróleo, Mexico.

References

1. A. W. Woods and S. J. Linz, *J. Fluid Mech.* **241**, 59–74 (1992).
2. O.M. Phillips, *Flow and reactions in permeable rocks* (Cambridge University Press, Cambridge, 1991).
3. E. J. Shaughnessy and J. W. Van Gilder, *Numer. Heat Transfer Part A* **28**, 389–408 (1995).
4. S. J. Linz and A. W. Woods, *Phys. Rev. A* **46**, 4869–4878 (1992).

BASIC NUMERICAL SIMULATIONS IN KALUZA'S MAGNETOHYDRODYNAMICS

ALFREDO SANDOVAL-VILLALBAZO, ANA LAURA GARCÍA-PERCIANTE

Departamento de Ciencias, Universidad Iberoamericana, Prolongación Paseo de la Reforma 880, Col. Lomas de Santa Fe, 01210 México, D.F., México
E-mail: alfredo.sandoval@uia.mx, adn229@onebox.com

LEOPOLDO S. GARCÍA COLÍN*

Departamento de Física, Universidad Autónoma Metropolitana, Av. Purísima y Michoacán S/N, Col. Vicentina, 09340 México, D.F., México
E-mail: lgcs@xanum.uam.mx

Relativistic magnetohydrodynamics is discussed within the framework of irreversible thermodynamics. This is done using Kaluza's ideas about unifying external fields in terms of the corresponding space–time curvatures for a given metric. The outcome of this approach is rewarding. The conservation equations follow in a direct way as well as the entropy balance equation with an entropy production whose form suggests the type of constitutive equations that are consistent with its semipositive definite property. Further, the resulting transport equations are of a hyperbolic type in agreement with causality. Specific numerical examples will be presented by way of illustration.

1 Introduction

The purpose of this work is to present a numerical solution to a heat transport equation formulated within the context of relativistic magnetohydrodynamics. Transport equations arise from the coupling of conservation or balance equations for a set of arbitrarily chosen state variables with the so called constitutive equations. These latter ones express a possible relationship between the currents that appear in the system when subjected to the action of external gradients with such gradients given in terms of the intensive variables. Thus a temperature gradient induces a heat (energy) current and so on. When this problem is formulated within the framework of classical irreversible thermodynamics, it is possible to establish a set of constitutive equations which guarantee that the entropy production is nonnegative, thus at grips with the second law of thermodynamics.

Here we wish to add two features to the above scheme. First, to formulate transport equations within the frame of general relativity. Second, to

---

*Also at El Colegio Nacional, Luis González Obregón 23, Centro Histórico 06020, México, D.F. México.

couple a fluid, assuming it is electrically charged, to an external electromagnetic field. The first objective, to derive relativistic transport equations in the absence of external fields has been previously discussed by two of us[1]. The main advantage of this formalism is not only to show its consistency with the second law of thermodynamics but to exhibit that the resulting transport equations are of the hyperbolic type, namely, they do not violate causality. The second objective, to extend this formalism to magnetohydrodynamics poses some challenges. When the problem is dealt with nonrelativistically and one introduces the electromagnetic field as an external force, the Lorentz force, two shortcomings arise[2]. Firstly, the resulting transport equations are of the parabolic type, violating causality, and secondly, in the entropy production there appears a bilinear term containing the electrical current times the electric field. This implies that to obey the second law one has to postulate Ohm's equation as a constitutive relation, a fact which in our opinion might seem too restrictive[3]. In view of these facts, the question is how to get around them to obtain the desired results. The answer is provided by the so called Kaluza–Klein formalism which, as we shall see, does provide a rather elegant and simple solution.

Indeed, in 1921 T. Kaluza[4-6] proposed a method to deal with a fundamental question in physics, the dualism of gravity and electricity. Quoting from the first lines of his paper, "Alongside the metric tensor of the 4-dimensional manifold (interpreted as a tensor potential of gravity) a general relativistic description of world phenomena requires also an electromagnetic four potential A_μ".

In order to accomplish such formulation, Kaluza proposed working in five dimensions to be able to add a fifth component to Einstein's stress tensor in a form which leads in a natural way to the covariant formulation of Maxwell's equations for the electromagnetic field. The essence of Kaluza's ideas and how we have used them to formulate relativistic magnetohydrodynamics will be discussed in the following section.

2 Relativistic magnetohydrodynamics

Following Kaluza we assume that the equations of general relativity may be applied to a 5×5 metric defined as,

$$g_{\mu\nu} = \begin{bmatrix} g_{11} & g_{12} & g_{13} & g_{14} & A_1 \\ g_{21} & g_{22} & g_{23} & g_{24} & A_2 \\ g_{31} & g_{32} & g_{33} & g_{34} & A_3 \\ g_{41} & g_{42} & g_{43} & g_{44} & \frac{1}{c}\varphi \\ A_1 & A_2 & A_3 & \frac{1}{c}\varphi & g_{55} \end{bmatrix} \tag{1}$$

To keep this presentation as simple as possible the components of the metric $g_{\mu\nu}$ will be taken to correspond to Minkowski's metric namely $g_{\mu\nu} = -\delta_{\mu\nu}$ except for $g_{44} = 1$. Also $[x^\mu]^t = \left[x^1, x^2, x^3, ct, x^5\right]$ and $[v^\mu]^t = \left[\dot{x}^1, \dot{x}^2, \dot{x}^3, c, \frac{q}{m}\right]$ where $[\;]^t$ denotes the transposed vector, $\dot{x}^i \equiv \frac{dx^i}{d\tau}$ where τ is the proper time and $v^5 = \frac{q}{m}$ where q and m are the charge and mass of the fluid's particles. This definition of v^5 guarantees that for small specific charges and negligible time dilation one recovers the well-known Lorentz equation of motion of a charged particle in the presence of the electromagnetic field.

The important results follow from Eq. (1) using Minkowski's metric by direct computation of the Christoffel symbols. The first one is the well known expression for the field tensor,

$$F^{\mu\nu} = \begin{bmatrix} 0 & B_z & -B_y & -\frac{1}{c}E_x \\ -B_z & 0 & B_x & -\frac{1}{c}E_y \\ B_y & -B_x & 0 & -\frac{1}{c}E_z \\ \frac{1}{c}E_x & \frac{1}{c}E_y & \frac{1}{c}E_z & 0 \end{bmatrix} \tag{2}$$

since $\mathbf{B} = \nabla \times \mathbf{A}$ and $\mathbf{E} = -\nabla\varphi + \dot{\mathbf{A}}$ as is well known. Moreover

$$F^{\lambda,\alpha\beta} + F^{\alpha,\beta\lambda} + F^{\beta,\lambda\alpha} = 0 \tag{3}$$

are precisely Maxwell's equations without sources, where $F^{\lambda,\cdots}$ denotes the covariant derivative and the remaining two equations are obtained through Einstein's field equation

$$G^{\mu\nu} = KT^{\mu\nu} \tag{4}$$

where $G^{\mu\nu}$ is Einstein's tensor, $T^{\mu\nu}$ is defined by

$$T^{\mu\nu} = \rho_0 v^\mu v^\nu - p_0 g^{\mu\nu} + \frac{p_0}{c^2} v^\mu v^\nu + \Xi^{\mu\nu} \tag{5}$$

and $T^{5\alpha} = T^{\alpha 5} = j_{el}^\alpha$ is the four dimensional electric density current vector. ρ_0 and p_0 are the local density and pressure in the commoving system and $\Xi^{\mu\nu}$ the viscous stress tensor. K is the gravitational constant.

It should be emphatically stressed that in all this formalism the Kaluza "cylinder condition" is required, namely, $\frac{\partial}{\partial x^5} = 0$. The only justification for it is that it leads directly to Maxwell's field equations in cartesian coordinates.

The second result is that the equation of motion of a particle along a geodesic is given by

$$\frac{d^2 x^\alpha}{ds^2} + \Gamma^\alpha_{\mu\nu} \frac{dx^\mu}{ds} \frac{dx^\nu}{ds} = 0 \tag{6}$$

where ds^2 is the line element and $\Gamma^\alpha_{\mu\nu}$ the ordinary Christoffel symbols. As said above, this reduces to Lorentz's equation in the nonrelativistic limit.

We now turn to magnetohydrodynamics omiting all algebraic steps which are identical in structure to those given in Sandoval-Villabazo and García-Colín,[1] remembering that here greek indices run from 1 to 5. Starting from Eq. (5) one uses the fundamental conservation equation

$$T^{\mu\nu}_{;\nu} = 0 \tag{7}$$

with $\frac{d}{d\tau} = v^\nu ()_{;\nu}$, $()_{;\nu}$ indicating a covariant derivative, to obtain the continuity equation for ρ_0 [7] and an internal energy balance equation[1,8] which we shall not bother to write down. The only assumption made in this derivation is that in the comoving system the total energy is conserved. The next step is crucial namely, we assume that the entropy flux is a 5D vector of the form

$$J^\nu_T = \rho_0 s_0 v^\nu + J^\nu \tag{8}$$

where the convective term $\rho_0 s_0 v^\nu$ containing the local entropy density s_0 defined in the commoving system yields the entropy flow through the hypersurface enclosing any arbitrary five dimensional volume and J^ν is the entropy flow generated by the dissipative processes taking place within that volume. Thus, the general form for the entropy balance equation is that

$$J^\nu_{T;\nu} = \sigma \geq 0 \tag{9}$$

where σ, the entropy production, according to the second law must be nonnegative. When combined with the continuity equation this leads to the general entropy balance equation and in particular, to the explicit form for σ namely,

$$\sigma = \frac{-J^\nu_q T_{0;\nu}}{T_0^2} - \frac{\Xi^{\mu\nu}}{T_0} \frac{\partial v_\mu}{\partial x_\nu} - \Gamma^\alpha_{\mu\nu} v_\alpha \frac{\Xi^{\mu\nu}}{T_0} \tag{10}$$

Here, J^ν_q is the five dimensional heat current and T_0 the local temperature defined in the comoving system. Eq. (10) is worth a minute of reflection. First of all, the only additional assumption required in its derivation besides the validity of the conservation equations in the comoving system is that in such system the *local equilibrium assumption* holds true. This assumption, stating that the local entropy density is a time independent functional of the conserved densities, ρ_0 and e_0 the internal energy density in this case, is the only assumption drawn from classical irreversible thermodynamics[9]. It allows one to write that if $s_0 = s_0\left(\rho_0, e_0\right)$,

$$\frac{ds_0}{d\tau} = \left(\frac{\partial s_0}{\partial \rho_0}\right)_{e_0} \frac{d\rho_0}{d\tau} + \left(\frac{\partial s_0}{\partial e_0}\right)_{\rho_0} \frac{de_0}{d\tau} \tag{11}$$

thus relating the dynamics of the local entropy to the one prescribed by the evolution equations. Eq. (11) is the underlying basis in the derivation of Eq. (10). Second of all, the form of σ is a bilinear one containing the sums of products of forces $T_{;\nu}$, $v^\mu_{;\nu}$, time fluxes, $\frac{J^\nu_q}{T^2_0}$, $\frac{\Xi^{\mu\nu}}{T_0}$. Thus, in the third term there is a force $\Gamma^\alpha_{\mu\nu} v_\alpha$ arising from the geometrical structure of our hyperspace through $\Gamma^\alpha_{\mu\nu}$ where the effects of the electromagnetic field are included. This is precisely what Kaluza proposed to achieve, to incorporate the effects of the electromagnetic field (or any other field) through the space curvature effects. One can indeed show[3] that in the absence of viscous forces this term reduces to the standard one $\left(-j_k E_k\right)$ of nonrelativistic magnetohydrodynamics. Finally, if one proposes constitutive equations relating the forces to the fluxes in a linear way, $J^\nu_q = -\lambda T_{;\nu}$, etc., σ will be non negative provided λ the thermal conductivity and the two other coefficients are nonnegative. This condition is borne by experiment in the nonrelativistic limit and we expect that it also holds true here.

The last step required is to obtain a heat transport equation from the formalism developed and this is a question of standard algebra. Using the local equilibrium assumption for the internal energy density, $e_0 = e_0\left(\rho_0, T_0\right)$ taking the time derivative with respect to the proper time, using the standard thermostatic formula for the coefficients $\left(\frac{\partial e_0}{\partial \rho_0}\right)_{T_0}$ and $\left(\frac{\partial e_0}{\partial T_0}\right)_{\rho_0} \equiv C_\rho$ and assuming valid the relativistic version of Fourier's equation, $J^\nu_q = -\lambda T_{;\nu}$ one gets that the equation for the local temperature is given by

$$\frac{2\lambda}{r}\frac{\partial T_0}{\partial r} + \frac{\partial^2 T_0}{\partial r^2} = f_1\left(r\right) + f_2\left(r\right) \tag{12}$$

where the source terms $f_1\left(r\right)$ and $f_2\left(r\right)$ are:

$$f_1\left(r\right) = \left(-\frac{GM^2}{c^2 r^3} + \frac{GM}{2r^2} + \frac{GMc^2}{2r\left(2GM - c^2 r\right)} - \frac{c^4 r}{4\left(2GM - c^2 r\right)} + \frac{3E_r v^5}{4}\right)\eta_r \tag{13}$$

$$f_1(r) = \left(-\frac{GM^2}{c^2 r^3} + \frac{GM}{2r^2} + \frac{GMc^2}{2r\left(2GM - c^2 r\right)} - \frac{c^4 r}{4\left(2GM - c^2 r\right)} + \frac{3E_r v^5}{4} \right) \eta_s$$

$$(14)$$

In equations (13) and (14), G is the universal gravitational constant, M the mass of the gravitational source and η_r and η_s the rotational and shear viscosities, respectively, and we have made use of the following simplifications:

1. Spherical symmetry without rotation.

2. Null magnetic field and radial electric field.

3. Small charge–mass ratio.

4. The local internal energy is assumed to be much smaller than the mechanical rest energy.

Eq. (12) is the resulting equation for the heat transport that will be used to obtain some numerical results for the case of a neutron star. Bulk viscosity does not appear in Eq. (12) because the divergence $v^\nu_{;\nu}$ vanishes. The main results are shown in Fig. 1, which shows the effects of the electric and gravitational fields on the temperature distribution of a dense sphere. The data corresponds to an astrophysical object of a solar mass, the radius of a neutron star, and temperature boundary values of the order of magnitude of this type of object. The intensity of the coupling of the electric field and the viscosities in the upper curve is of the order of 10^4 W m^{-2}.

Due to the fact that this formalism is completely new in the context of magnetohydrodynamics[3], we expect that powerful numerical codes may be developed in the future in order to get a deeper understanding of this Kaluza's five-dimensonal tool. Although the final equation analyzed here included many simplifications that lead to a solution easily achieved by very basic algorithms (Mathematica's NDSolve), the main body of the formalism involves a very complex new system of transport equations. We therefore leave this question to experts in numerical methods to tackle it.

Concluding, this example simply ilustrates the versatility of the fomalism which may be extended to much more realistic and thus, complicated situations. Obtention of results is simply a question of powerful numerical computations.

Details concerning the numerical procedure of the results here quoted are too cumbersome to be presented in such a short paper. However, they are available to any interested reader by contacting the first two authors.

254

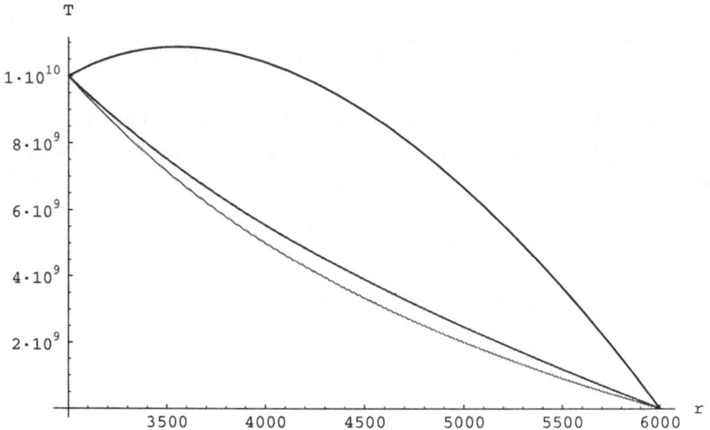

Figure 1. The graph above shows the contrast of temperature distribution in three cases. Horizontal axes corresponds to distance from the origin (in meters). The vertical axis represents temperature (in Kelvins). The lower curve was calculated in the absence of both fields. The middle curve shows the effect of the gravitational field, while the Kaluza's effects of an intense electric field cause the existence of a maximum of temperature, as shown in the upper curve.

References

1. A. Sandoval-Villalbazo and L. S. García-Colín, *Physica* A **234**, 358–370 (1996); *ibid.* A **240**, 480 (1997).

2. S. R. de Groot and P. Mazur, *Non-equilibrium Thermodynamics* (Dover Publications, Mincola, N.Y., 1984)

3. A. Sandoval-Villalbazo and L. S. García-Colín, *Phys. of Plasmas* **7**, 4823–4830, (2000).

4. Th. Kaluza, *Sitzungsberg. D. Preuss. Akademie D. Wissenschaften, Physik.-Mathemat. Klasse*, 966–972 (1921).

5. *Modern Kaluza–Klein Theories*, T. Appelquist, A. Chodos and P. Freund, eds., (Addison-Wesley, Reading, Mass., 1987).

6. L. O'Raifeartaigh and N. Strauman, *Rev. Mod. Phys.* **72**, 1 (2000).

7. A. Sandoval-Villalbazo and L. S. García-Colín, *J. Non-Equilib. Thermodyn.* **24**, 380–386 (1999).

8. A. Sandoval-Villalbazo and L. S. García-Colín, *Gen. Rel. and Grav.* **31**, 781 (1999).

9. L. S. García-Colín and F. J. Uribe, *J. Non-Equilib. Thermodyn.* **16**, 89 (1991).

UNDERSTANDING DILUTE GASES: GOING BEYOND THE NAVIER–STOKES EQUATIONS

F. J. URIBE

Departamento de Física, Universidad Autónoma Metropolitana–Iztapalapa, 09340 México, D.F., México

The main interest of this work is the Chapman–Enskog method for solving the Boltzmann equation. The calculation of the diffusion coefficient using the classical trajectory method for three realistic potentials for He–N_2 is reported. A comparison of the calculated diffusion coefficients with experimental data and the extended corresponding states principle is also performed. Limitations of the Navier–Stokes equations for some problems are mentioned and a more detailed analysis of this point for the shock wave problem is considered.

1 Introduction

The consideration of dilute gases trades wealth of applications for depth of understanding. In this case the Boltzmann equation, or many of its variants for dilute gases with structure either quantum or classical, provides a sound theoretical description[1]. In this work I will discuss the Chapman–Enskog method[2], Grad's method[3] and the Direct Simulation Monte Carlo (DSMC) method[4] for solving the Boltzmann equation.

The Chapman–Enskog method solves the Boltzmann equation by making an expansion of the distribution funcion in terms of the Knudsen number; to order zero the heat flux and pressure tensor do not contain dissipative phenomena and substitution of both quantities in the conservation equations result in the Euler equations of hydrodynamics. Dissipative effects are incorporated to first order in the Knudsen regime through the thermal conductivity and the shear viscosity. The well known Navier–Stokes–Fourier linear constitutive equations are obtained and the transport coefficients can be calculated for a given intermolecular potential. Substitution of the constitutive equations in the conservation equations gives rise to the well known Navier–Stokes equations which have many applications. To second order in the Knudsen regime the Chapman–Enskog method gives rise to the Burnett equations, which are also considered in this work. Traditionally the problem of solving the hydrodynamic problem for a given set of boundary conditions, the Navier–Stokes equations with the no-slip at the boundary conditions for example, was considered outside of the scope of the kinetic theory. With the advent of the DSMC method, which is now feasible with current computer resources, it is possible to obtain information not only about the hydrodynamic fields but also about the distribution function for which experimental information is also available[5]. Furthermore, Nonequilibrium Molecular Dynamics (NEMD) simulations for shock waves in dilute gases are now possible and the results are in agreement with those of DSMC[6]. Since there is also a validation of the DSMC method using experimental data[7], I decided to use the implementation by Bird[4] in the shock wave problem.

The structure of this work is as follows. After this introduction I discuss the evaluation of transport properties at the Navier–Stokes regime, the main objective being to give a rough idea of the level of precision that the kinetic theory has reached. In section 3 some examples are shown in which the hydro-dynamic description provided by the Navier–Stokes equations is susceptible of improvement and the case of shock waves is considered in some detail to illustrate this point. Finally, some closing remarks are given.

2 The Navier–Stokes regime

Before tackling the Burnett equations I would like to discuss some applications at the Navier–Stokes regime in order to get an idea of the level of refinement that the kinetic theory has reached. One method that allows to calculate transport properties, for a given intermolecular potential, is the so called classical trajectory method[8]. This method solves the differential equations of motion for a binary encounter under different initial conditions. From these solutions, a number of effective cross sections can be calculated and the transport coefficients are then accordingly obtained. A natural question arises: what is the true interaction potential for a given system? The field of intermolecular potentials is vast and sophisticated[9] and here I will consider the calculation of the diffusion coefficient for three intermolecular potentials for He–N_2 using the classical trajectory method. There is another route to calculate transport properties in which one does not pay attention to the explicit form of the intermolecular potential, but rather assume that its form is the same for a class of systems. This route is known as the extended corresponding state principle and a review of its present status is available[10].

In figure 1 the percentage deviations of the diffusion coefficient (Δ_D^0) with respect to the extended corresponding state principle as a function of the temperature is given. Δ_D^0 is defined as,

$$\Delta_D^0 \equiv \frac{\left(D_{12}^{\text{theo,exp}} - D_{12}^{\text{cs}} \right)}{D_{12}^{\text{cs}}} \times 100, \tag{1}$$

where $D_{12}^{\text{theo,exp}}$ are the theoretical or experimental values of the diffusion co-

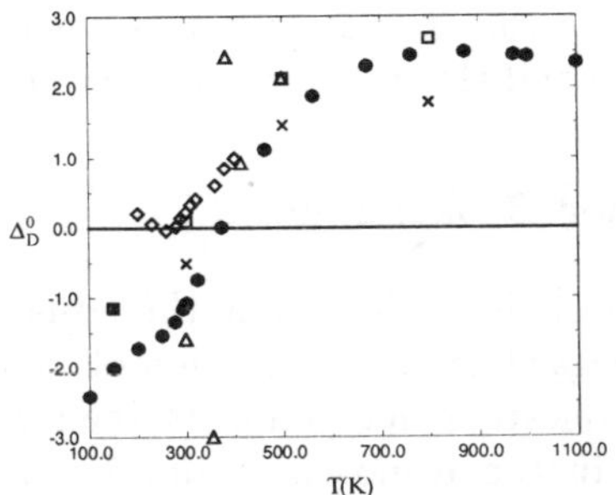

Figure 1. Percent deviation (Δ_D^0) of experimental and theoretical data with respect to the extended corresponding states principle as a function of the temperature. Crosses: values for the BTT potential; solid circles: values for the potential by Hu and Thakkar (present work); squares: values for the 3SMV potential; triangles: experimental values by Seager *et al.*; diamonds: experimental values by Trengove *et al.*

efficient and D_{12}^{cs} is the value for the diffusion coefficient given by the extended corresponding states principle for a mole fraction of N_2 equal to 10^{-6}. The two sets of experimental data shown, classified as primary data by Bzowsky *et al.*[16], are the ones by Tengrove *et al.*[11] with an estimated accuracy of about 0.5 % and by Seager *et al.*[12] which have lower accuracy and do not correspond to the limit of a trace component of N_2 as the ones by Tengrove *et. al.* The three potential energy surfaces (PES) considered in the figure are the semiempirical model by Bowers *at al.*[13] (BTT), the multiproperty fit surface by Gianturco *et al.*[14] (M3SV) and the potential obtained by Hu and Thakkar[15] which is an analytical fit to *ab initio* results obtained with the supermolecule fourth-order Møller–Plesset perturbation theory. The effective cross sections for the BTT and M3SV potentials were taken from the work by Dickinson and Heck[8], the effective cross sections for the potential by Hu and Thakkar were obtained in collaboration with Prof. A. S. Dickinson of the University of Newcastle upon Tyne (unpublished).

It can be concluded from Fig. 1 that the difference between the best experimental values (those of Tengrove *et al.*), the calculated diffusion coefficients for three realistic intermolecular potentials, and the extended corresponding states principle, are within 3% in the temperature range from 100 K to 1000 K. In my opinion such an agreement is remarkable. Other transport properties can also be considered but a more extensive analysis is beyond the objective of this work. While the calculation of some transport properties is in good shape, one must not infer that the same is true for predictions given by the Navier–Stokes equations and this point will be considered in the next section.

3 Beyond the Navier–Stokes regime

From the previous section it is clear that the situation at the level of the Navier–Stokes regime regarding the calculation of some transport coefficients is satisfactory, and the question that jumps to the fore is, why it is necessary to go to second order in the Knudsen number? The reason is that linear constitutive equations are expected to hold for small gradients and that there are a number of interesting problems for which this condition does not hold; a well known example corresponds to the propagation of shock waves[17]. For plane Poiseuille flow, where the flow is subsonic and the gradients are not very large, Uribe and Garcia[18] solved the autonomous six dimensional first-order dynamical system provided by the Burnett equations as an initial value problem using Adams method, the DSMC data provided the initial conditions. They found a nonconstant pressure profile and a heat flow along the direction of the flow, in agreement with the DSMC simulations, such features cannot be explained with the Navier–Stokes equations. A further example in which the Burnett equations improve on the Navier–Stokes equations is the case of a one dimensional strongly nonisothermal gas[19]. I do not have space to do justice to the above-mentioned and other works, and in the following I will concentrate on certain aspects of the shock wave problem.

3.1 Shock waves

The situation with shock waves in monatomic gases is somewhat puzzling. First of all let me point out that for a Mach number of value 100 the temperature at the hot side of the shock, assuming standard conditions at the cold part of the shock, is such that the system must be ionized, so the problem of finding structure for high Mach numbers with the Boltzmann equation is academic[20]. Also, the Navier–Stokes equations predict that for some models and any Mach number (M) there exists structure. Grad's thirteen moment approximation, which in principle provide us with a method for going beyond the Navier–Stokes regime, give structure for Mach numbers lower than 1.65. This fact led Weiss[21] to point out that although the Navier–Stokes equations are not very good to predict shock wave profiles for large Mach numbers it turns out that Grad's thirteen moment approximation is even worse. Weiss' remark reflects the point of view that a theory is not good if it cannot provide structure for all Mach numbers, but in the opinion of some authors[20] it is necessary to know first if the Boltzmann equation has structure for all Mach numbers, before the argument can be raised against a theory. One could think that the problem of finding structure for larger Mach numbers can be solved by considering more moments, but there is work by Holway[22] who claimed that if more moments are considered there is always a bound on Mach numbers ($M \approx 1.85$) such that no structure can be found for Mach numbers larger than the bound. Weiss[23] recently challenged the work by Holway.

The Navier–Stokes equations provide structure for all Mach numbers[24] and the situation regarding their results is again controversial[25]. For some authors these equations provide a poor description even for Mach numbers close to one and for others they provide an accurate description in this case[25]. The explicit stationary Navier–Stokes equations for the case of rigid spheres are the following[22],

$$f \, u^{\star\prime}(s) = \tau_0 + 1 - u^\star(s) - \frac{\tau(s)}{u^\star(s)}, \tag{2}$$

$$g \, \tau'(s) = \frac{3}{2}\tau_0 + \frac{1}{2}(1 - u^\star(s))^2 + \tau_0(1 - u^\star(s)) - \frac{3}{2}\tau(s), \tag{3}$$

where $u^\star(s) = u(x)/u_0$, $u(x)$ is the stationary velocity as a function of the position, u_0 is the velocity at the cold part of the shock, $s = x/l$ where l is the "mean free path" as defined by Holian $et\ al.$[26], $\tau(s) = k\,T(x)/m\,u_0^2$ is the reduced temperature, k the Boltzmann constant, $T(x)$ the temperature, m the mass, $\tau_0 = \sqrt{3/(5\,M)}$ the reduced temperature at the cold part of the shock and the prime denotes the derivative with respect to s. For the rigid sphere model $f = -\sqrt{\tau(s)}$ and $g = -\frac{45}{16}\sqrt{\tau(s)}$. Equations (2) and (3) provide a dynamical system with well known properties, the upstream critical point is an unstable node, the downstream critical point is a saddle and there is a unique curve joining these two critical points (heteroclinic trajectory). Futhermore, if upstream corresponds to $s = -\infty$ and downstream to $s = \infty$, then perturbing slightly the downstream critical point and performing the integration in the negative direction, the solution will go to the upstream critical point. This method of approximately obtain the structure for the shock wave problem goes back to Gilbarg and Paolucci[24]. A numerical analysis of the Navier–Stokes dynamical system using Adams method, the Runge–Kutta method and the backward differentiation formula (BDF), is available[27]. The BDF and Adams method have also been used to solve the Burnett equations[20,28].

The idea of perturbing downstream and integrating towards upstream was used by Sherman and Talbot[29] to numerically solve the stationay Burnett equations, which constitute a first-order autonomous dynamical system in four dimensions, for Maxwellian molecules. They found that the solutions developed large oscillations and concluded that there was little merit in the Burnett equations. Later on, Foch and Simon[30] argued that before these oscillations develop the numerical solution is very near upstream and in this way they reported solutions up to $M = 4$ for Maxwellian molecules. The situation regarding the local structure for the Burnett equations was considered by Foch[31] and he found that for Maxwellian molecules there exists a bifurcation at upstream in which the upstream unstable node changes to a saddle, see also [32,33]. Similar conclusions regarding the local behavior around the critical points for the rigid sphere model have been reported[28]. Simon and Foch also brought attention to the work by Montgomery[34] regarding the existence of an heteroclinic trajectory (but not uniqueness) for the Burnett equations. Montgomery's theorem is actually of more general validity and can be used to investigate the existence of higher-order equations, for example the super-Burnett equations. The above mentioned complications of the Burnett equations prompted other authors to investigate other approaches. Fiscko and Chapman[35] tried a nonstationary code and were able to obtain numerical solutions up to $M = 50$ for the Burnett and super-Burnett equations using different models. They compared the results with experimental data and their implementation of the DSMC method. However, apart from the fact that their DSMC calculations were at fault, they discovered an anomaly of the Burnett equations that consisted in numerical solutions being unstable for high Mach numbers or when a fine computational mesh was used[36]. Another direction that has been explored is the one by Reese[33] in which the boundary conditions for the stationary Burnett equations are modified, he reported results for shock waves in argon up to $M = 8$.

In 1992 Salomons and Mareschal[6] reported the first Molecular Dynamics (MD) calculations for a dilute gas of rigid spheres and showed that the fluxes given by the Burnett equations were in better agreement, with respect to the MD calculations, than the Navier–Stokes equations for Mach numbers about 100. Shortly after this report Salomons and Mareschal in collaboration with Holian and Patterson[26] tested a conjecture due to Holian; they found that the conjecture provided a more accurate description for shock wave profiles than the Navier–Stokes equations. The dynamical system obtained by Holian $et\ al.$[26] results from taking $f = -\sqrt{u^\star(s)\left(\tau_0 + 1 - u^\star(s)\right)}$ and $g = -\frac{45}{16}\sqrt{u^\star(s)\left(\tau_0 + 1 - u^\star(s)\right)}$ in Eqs. (2) and (3). Since the eigenvalues of the Jacobian matrix at the critical points are real, for both the Navier–Stokes equations and the theory advanced by Holian $et\ al.$, oscillatory solutions are not expected. Dividing Eq. (3) by Eq. (2) it follows that the derivative of τ with respect to $u^\star$ is the same for the Navier–Stokes equations and the Holian dynamical system; this means that while the shock profiles are different the orbits are the same. So, in the $u^\star$-τ plane there is no difference between the Navier–Stokes and Holian dynamical systems. Recent numerical calculations of the orbits for both theories corroborate this fact[20].

After the work by Salomons and Mareschal it was a natural step to see is one could provide stationary shock wave profiles for high Mach numbers. In 1995 López de Haro and Garzó[37] reported shock wave profiles for high Mach numbers for the linearized Burnett equations in a dense gas of rigid spheres and in 1998 profiles for the full Burnett equations for a dilute gas of rigid spheres at $M = \infty$ were reported[28]. However, for the dilute case it was mentioned[28] that the calculated orbit did not corresponds to an heteroclinic trajectory. Furthemore, the calculated orbit does not corresponds to an heteroclinic trajectory for Mach numbers greater than $M_c \approx 2.69$ where the bifurcation at which the upstream unstable node changes to a saddle appears. The reason for this is the existence of a limit cycle for $M \in [M_c, M_{c_1}]$, $M_{c_1} \approx 3.25$, and for larger Mach numbers than M_{c_1} to the fact that the local flow can not be extended to $s = -\infty$. I will refer to the bifurcation at M_c as a saddle-node–Hopf bifurcation. For Mach numbers lower than M_c numerical evidence suggest that the calculated orbit corresponds to an heteroclinic trajectory, Montgomery's theorem can shown to hold for the Burnett equations[20] but the theorem does not give information about the explicit range of Mach numbers in which the heteroclinic trajectory exists. I would like to stress that althought for $M > M_c$ the calculated orbit does not corresponds to an heteroclinic trajectory, this does not mean that the Burnett equations do not have structure. In Fig. 2 the projection of the calculated orbit for the Burnett dynamical system in the $u^\star$-τ plane at $M = 2.68$ is given, the orbits for the Navier–Stokes equations at the same Mach number can also be found. Comparisons in other planes at $M = 2$ are also available[20].

In 1982 Bobylev[38] showed that for a dilute gas of Maxwellian molecules which is at rest, the Burnett equations are unstable in the hydrodynamic sense. His result relied on the use of linear stability analysis and it was used by Zhong *et al.*[40] to explain the anomaly found by Fiscko and Chapman. Recently[39], Bobylev's instability has been considered for rigid spheres and it has been argued that it can be interpreted as to give a bound on the Knudsen number above which the Burnett equations are not valid. The Chapman–Enskog expansion is valid for small Knudsen numbers but an explicit range has never been given, it would be interesting to see if the anomaly found by Fiscko and Chapman can be recast in terms of the Knudsen number. The Stanford group[36,40] has considered the stabilization of the Burnett equations; this has been done by dropping some terms in the Burnett equations or by including in them terms that come from the super-Burnett equations. As far

262

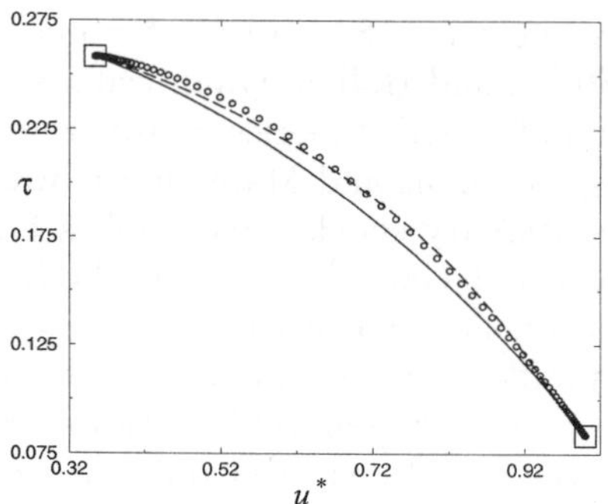

Figure 2. Orbits in the u^*-τ plane at $M = 2.68$ for rigid spheres. Circles: DSMC; solid line: Navier–Stokes; dashed line: Burnett; squares: critical points.

as I know it is not known if their subset or superset of the Burnett equations is also satisfactory in situations different from the shock wave problem. Another modification to the Burnett equations due to Woods[41] has been used to obtain shock wave profiles for Mach numbers up to $M = 30$, satisfactory results have been reported when compared with experimental data for argon[42].

To end this section I would like to draw some conclusions, (a) the results shown in Fig. 2 show that the Burnett equations provide an improvement over the Navier–Stokes equations for small Mach numbers, (b) it is simpler and better to compare the underlying dynamical systems instead of only the profiles, (c) the orbits of the Navier–Stokes and Holian dynamical systems are the same, (d) the calculated orbit and profiles of the Burnett equations for strong shocks, $M > 100$, improve on the Navier–Stokes equations, but the results are not conclusive due to the lack of structure for the calculated orbit; it would be interesting to know if the Boltzmann equation provides structure for such Mach numbers; (e) Bobylev's instability can be interpreted as providing a bound for the Mach number above which the Burnett equations are not valid, and it would be interesting to see if the anomaly found by Fiscko and Chapman can be interpreted in terms of the Knudsen number; (f) the modifications of the Burnett equations provided by the Stanford and Oxford groups should be tested for situations different from the shock wave problem.

4 Final remarks

The Burnett equations provide a better description than the Navier–Stokes equations in some examples; the main problem with them is that the boundary conditions are unknown.

Moment methods provide another alternative for extending the Navier–Stokes equations but the closure problem remains an open task to solve.

Acknowledgments

I thank R. M. Velasco for a critical reading of the manuscript. Work financed in part by CONACyT.

References

1. F. R. W. McCourt, J. J. M. Beenakker, W. E. Köhler and I Kuščer, *Nonequilibrium Phenomena in Polyatomic Gases* (Oxford University Press, New York, 1990).
2. S. Chapman & T. G. Cowling, *The Mathematical Theory of Non-Uniform Gases* (Cambridge University Press, 1970).
3. H. Grad, *Comm. Pure and Appl. Math.* **2**, 331 (1949).
4. G. A. Bird, *Molecular Gas Dynamics and the Direct Simulation of Gas Flows* (Clarendon, Oxford, 1994)
5. G. C. Pham-Van-Diep, D. A. Erwin and E. P. Muntz, *Science* **245**, 269 (1989).
6. E. Salomons and M. Mareschal, *Phys. Rev. Lett.* **69**, 269 (1992).
7. G. C. Pham-Van-Diep, D. A. Erwin and E. P. Muntz, *Phys. Fluids A* **3**, 697 (1991).
8. A. S. Dickinson and E. L. Heck, *Molec. Phys.* **70**, 239 (1990).
9. A. van der Avoird, in *Status and Future Developments in the Study of Transport Properties*, W. A. Wakeham, A. S. Dickinson, F. R. W. McCourt and V. Vesovic, eds. (Kluwer, 1992).
10. E. A. Mason and F. J. Uribe, in *Transport Properties of Fluids*, J. Millat, J. H. Dymond and C. A. Nieto de Castro, eds. (Cambridge University Press, New York, 1996).
11. R. D. Tengrove and P. J. Dunlop, *Physica A* **115** , 339 (1982).
12. S. L. Seager, L. R. Geertson and J. C. Giddings, *J. Chem. Eng. Data* **8**, 168 (1963).
13. M. S. Bowers, K. T. Tang and J. P. Toennies, *J. Chem. Phys.* **88**, 5465 (1988).
14. F. A. Gianturco, M. Venanzi, R. Candori, F. Pirani, F. Vecchiocattivi, A. S. Dickinson and M. S. Lee, *Chem. Phys.* **109**, 417 (1986); *ibid* **113**, 166 (1987).
15. C.-H. Hu and A. J. Thakkar, *J. Chem. Phys.* **104**, 2541 (1996).
16. J. Bzowski, J. Kestin, E. A. Mason and F. J. Uribe, *J. Phys. Chem. Ref. Data* **19**, 1179 (1990).
17. H. Alsmeyer, *J. Fluid. Mech.* **74**, 497 (1976).
18. F. J. Uribe and A. L. Garcia, *Phys. Rev. E* **60**, 4063 (1999).

19. D. W. Mackowski, D. H. Papadopoulos and D. E. Rosner, *Phys. Fluids* **11**, 2108 (1999).

20. F. J. Uribe, R. M. Velasco, L. S. García-Colín and E. Díaz-Herrera, *Phys. Rev. E* **62**, 6648 (2000).

21. W. Weiss, *Phys. Rev. E* **52**, R5760 (1996).

22. L. H. Holway, *Phys. Fluids* **7**, 911 (1964).

23. W. Weiss, *Phys. Fluids* **8**, 1659 (1996).

24. D. Gilbarg and D. Paolucci, *J. Ration. Mech. Anal.* **2**, 617 (1953).

25. G. C. Pham-Van-Diep, D. A. Erwin and E. P. Muntz, *J. Fluid Mech.* **232**, 403 (1991).

26. B. L. Holian, C. W. Patterson, M. Mareschal and E. Salomons, *Phys. Rev. E* **47**, R24 (1993).

27. F. J. Uribe, R. M. Velasco and L. S. García-Colín, *Phys. Rev. E* **62**, 3209 (1998).

28. F. J. Uribe, R. M. Velasco and L. S. García-Colín, *Phys. Rev. Lett.* **81**, 2044 (1998).

29. F. S. Sherman and L. Talbot, in *Proceedings of the First International Symposium on Rarefied Gas Dynamics*, F. M. Devienne, ed. (Pergamon, New York, 1960).

30. J. D. Foch Jr. and C. E. Simon, in *Rarefied Gas Dynamics*, J. L. Potter, ed., *Progress in Astronautics and Aeronautics* **51** (AIAA, New York, 1977).

31. J. D. Foch Jr, *Acta Physica Austriaca* suppl. X, 123 (1973).

32. J. M. Reese, F. Thiele and L. C. Woods, in *International Symposium on Rarefied Gas Dynamics*, J. Harvey and G. Lord, eds. (Oxford University Press, Oxford, 1995).

33. J. M. Reese, in *Proceedings of the 21st International Symposium on Shock Waves*, Paper 8050, A. F. P. Houwing, ed. (Panther Publishing and Printing, Canberra, 1997).

34. J. T. Montgomery, *Phys. Fluids* **18**, 148 (1975).

35. K. A. Fiscko and D. R. Chapman, in *Rarefied Gas Dynamics*, E. P. Muntz, D. P. Weaver and D. H. Campbell, eds. **118** (AIAA, Washington, 1989).

36. F. E. Lumpkin III and R. D. Chapman, *J. Thermophys. Heat Transfer* **6**, 419 (1992).

37. M. López de Haro and V. Garzó, *Phys. Rev. E* **52**, 5688 (1995).

38. A. V. Bobylev, *Soviet Phys. Dokl.* **27**, 29 (1982).

39. F. J. Uribe, R. M. Velasco and L. S. García-Colín, *Phys. Rev. E* **62**, 6648 (2000).

40. X. Zhong, R. W. MacCormark and D. R. Chapman, *AIAA Journal* **31**, 1036 (1993).

41. L. C. Woods, *An Introduction to the Kinetic Theory of Gases and Magnetoplasmas* (Oxford University Press, New York, 1993).

42. J. M. Reese, L. C. Woods, F. J. P. Thivet and S. M. Candel, *J. Comp. Phys.* **117**, 240 (1995).